Palgrave Studies in Crime, Media and Culture

This series aims to publish high quality interdisciplinary scholarship for research into crime, media and culture. As images of crime, harm and punishment proliferate across new and old media, there is a growing recognition that criminology needs to rethink its relations with the ascendant power of spectacle. This international book series aims to break down the often rigid and increasingly hardened boundaries of mainstream criminology, media and communication studies, and cultural studies. In a late modern world where reality TV takes viewers into cop cars and carceral spaces, game shows routinely feature shame and suffering, teenagers post 'happy slapping' videos on YouTube, both cyber bullying and 'justice for' campaigns are mainstays of social media, and insurrectionist groups compile footage of suicide bomb attacks for circulation on the Internet, it is clear that images of crime and control play a powerful role in shaping social practices. It is vital then that we become versed in the diverse ways that crime and punishment are represented in an era of global interconnectedness, not least since the very reach of global media networks is now unparalleled.

Palgrave Studies in Crime, Media and Culture emerges from a call to rethink the manner in which images are reshaping the world and criminology as a project. The mobility, malleability, banality, speed, and scale of images and their distribution demand that we engage both old and new theories and methods and pursue a refinement of concepts and tools, as well as innovative new ones, to tackle questions of crime, harm, culture, and control. Keywords like image, iconography, information flows, the counter-visual, and 'social' media, as well as the continuing relevance of the markers, signs, and inscriptions of gender, race, sexuality, and class in cultural contests, mark the contours of the crime, media and culture nexus.

Avi Brisman

Direct Action as Conceptual Art?

An Examination of the Role of the Communiqué for Ecodefense

palgrave
macmillan

Avi Brisman
School of Justice Studies
College of Justice and Safety
Eastern Kentucky University
Richmond, KY, USA

ISSN 2946-3912 ISSN 2946-3920 (electronic)
Palgrave Studies in Crime, Media and Culture
ISBN 978-3-031-76633-6 ISBN 978-3-031-76634-3 (eBook)
https://doi.org/10.1007/978-3-031-76634-3

Cover illustration: Lina Tharsing, *Controlled Burn*, oil on canvas, 12" x 9", 2024

This Palgrave Macmillan imprint is published by the registered company Springer Nature Switzerland AG
The registered company address is: Gewerbestrasse 11, 6330 Cham, Switzerland

For Nigel

About This Book

Ecodefense—which includes direct action in defense of or on behalf of the environment and nonhuman animals (e.g., economic sabotage, property destruction, lab raids, "live liberations")—attempts to reduce or stop immediately an environmental harm, risk, or threat. It stands in contrast to indirect action, which relies on representative democracy to bring about change. In some instances, an act of ecodefense is accompanied by a communiqué that, *inter alia*, explains the reason for the (specific) action and the broader purpose and goals of the activists (or ecodefenders). Whereas some consider the communiqué as inconsequential or secondary to the act of ecodefense (or as indistinguishable from education—a type of indirect action), others regard the communiqué as part of the act of ecodefense and vital to its ability to effect change.

Conceptual art is art in which the concept(s) or idea(s) involved in the work take precedence over traditional aesthetic and material concerns. Conceptual artists—many of whom privilege the planning and decisions over the execution—frequently document their work with photographs and written texts, but they insist that the "real" work of art has already occurred (i.e., that the work exists apart from the documentation in photographic and written form).

Direct Action as Conceptual Art?: An Examination of the Role of the Communiqué for Ecodefense explores the significance of the communiqué for ecodefense by comparing the communiqué to written texts in Conceptual art. The hope is that considering the communiqué in this light will help us better understand the reasons and rationales for so-called "radical environmentalism" undertaken for the purposes of avoiding, disrupting, eliminating, preventing, or reducing environmental harms, including natural resource exploitation and animal abuse. This short book also contemplates the extent to which ecodefense could and/or should serve to educate and generate support for environmental causes—especially those seeking to address the climate crisis.

Preface and Acknowledgments

I want to begin *Direct Action as Conceptual Art?: An Examination of the Role of the Communiqué for Ecodefense* by quoting from a work of the venerable Professor Rob Watts. Based on the title of my book, one might be inclined to surmise that I wish to draw on Professor Watts' *Criminalizing Dissent: The Liberal State and the Problem of Legitimacy* (2020). That would be a good guess, for *Criminalizing Dissent: The Liberal State and the Problem of Legitimacy* (2020) offers readers a rich account of diverse forms of dissent and the increasing ways in which liberal-democratic states have expanded the criminalization of (this) dissent.

Instead, I want to make reference to Professor Watts' *States of Violence and the Civilising Process: On Criminology and State Crime* (2016), which I had the pleasure of reading in manuscript form, shortly after meeting Professor Watts for the first time at a conference held in July 2015 at Queensland University of Technology in Brisbane, Queensland, Australia. In the opening sentence of the Preface to *States of Violence and the Civilising Process: On Criminology and State Crime* (2016: v), Professor Watts states that his book "had an unnaturally long gestation." He then proceeds to outline some of his intellectual realizations that delayed—although Professor Watts contends that these discoveries *strengthened*—his monograph. Ever self-reflective, Professor Watts adds that "many far more mundane inertial factors combined to ensure that this book [underwent] a much longer . . . gestation that [he] had thought either likely or desirable" (2016: viii). "One such [reason]," Professor Watts (2016: viii) continues, "was the lack of time. I neither sought nor received any research grants in order to investigate or write the book.Like so many other people who work in a university I did not need money ... just time."

Like *States of Violence and the Civilising Process: On Criminology and State Crime* (2016), my short book—really an essay—*Direct Action as Conceptual Art?: An Examination of the Role of the Communiqué for Ecodefense* experienced a long gestation. There were a number of reasons for this.[1] "Lack of time" is certainly one of them. I am employed by a regional comprehensive university in a state with one of the highest poverty rates in the country. Although I appreciate the occasion to instruct and learn with and from a population of economically disadvantaged, first-generation students (many of whom are working adults with families), my teaching load has increased over the years, while the opportunities for "course releases"/"reassigned time" (for research and scholarship) have effectively evaporated due, in part, to shrinking state budget allocations to higher education—a travesty and tragedy not peculiar to the Commonwealth of Kentucky. To be sure, I am not alone in this regard. As I am certain is the case with many professors at my institution and comparable ones across the United States, I feel grateful for the opportunity to participate in an educational endeavor that promotes upward mobility for underrepresented, low-income individuals. But this pedagogical responsibility has meant fewer avenues for research and scholarship.

Allocation of effort (or distribution of effort) aside, other factors contributed to this book's long gestation. *Direct Action as Conceptual Art?: An Examination of the Role of the Communiqué for Ecodefense* began as a paper, which I presented (in its embryonic stage) at the 61st Annual Meeting of the American Society of Criminology in Philadelphia, Pennsylvania, on November 5, 2009, in a thematic session entitled "Recent Studies in Green Criminology." As I delivered my paper, I was acutely aware of the presence in the audience of Emeritus Professor (then-Pro-Vice Chancellor and Professor) Nigel South of the University of Essex. The previous year, I had met "Nigel," as I know him now—but back then, he was "Professor South"—at the 60th Annual Meeting of the American Society of Criminology (in St. Louis, Missouri). Although Nigel and I had corresponded a few times between November 2008 and November 2009, I was still starstruck and eager to impress him. Nigel sat through my presentation with his arms folded in front of him, looking rather displeased. He held that pose for the three papers that followed mine (or so I seem to recall) and said nothing during the fifteen minutes or so devoted to questions and comments at the end of the eighty-minute session. (I had the misfortune of speaking first.) I did not encounter Nigel during the remaining days of the annual meeting, but his reaction—or what I assumed to be his reaction—haunted me. Thus, rather than return home at the conclusion of the annual meeting and transform the paper into an article for submission to a peer-reviewed journal, I put the piece aside and focused on other matters, with the goal of returning to it before too much time had elapsed.

It was not long before Nigel did reach out to me—but it had nothing to do with my paper, which at the time was called "Direct Action as Conceptual Art: An Examination of the Role of the Communiqué for Eco-Defense and Animal Liberation." (I have no idea when I inserted the question mark into the title and reconceptualized "ecodefense" as including "animal liberation.") Instead, Nigel wrote to me to inquire whether I might be willing to join him as an editor for the *Routledge International Handbook of Green Criminology* (2013). I was overjoyed: it was a career-defining moment and I still remember receiving it on a cloudy day, while sitting in my apartment on the thirty-second floor of my building on the Upper East Side of New York City. I was also fairly certain that Nigel had either forgotten about my paper on direct action as conceptual art or had chosen to overlook it as academic immaturity. Suffice it to say, I did not bring it up until years later—and after many successful (and ongoing!) collaborations with Nigel. When I finally did share this story with Nigel, he explained that he was not at all cross or unconvinced; rather, he said, he imagined that he was either "deep in thought," pondering what I was suggesting, or simply "acting very British" (presumably a reference to his facial expression and the position of his arms folded in front of him). Whatever the case may be, I have been most fortunate to have been able to learn from and work so closely with Nigel—now Emeritus Professor South—these last fifteen years. Had he not "distracted" me with so many exciting projects and offers to co-author and co-edit, my ideas about drawing comparisons between direct action and conceptual art might have reached the light of day much earlier. Of course, had Nigel not been so supportive and encouraging in all facets of my academic/professional and personal life—and had he not been so instrumental in the development of my thinking—perhaps the nascent paper, "Direct Action as Conceptual Art: An Examination of the Role of the Communiqué for Eco-Defense and Animal Liberation," would not have become this book: *Direct Action as Conceptual Art?: An Examination of the Role of the Communiqué for Ecodefense*. It is for this reason—and many others—that I dedicate this book to him.

As this exploration (now called *Direct Action as Conceptual Art?: An Examination of the Role of the Communiqué for Ecodefense*) lay dormant over the years—although I prefer to think of it as "simmering"—I never abandoned the belief that I would complete it. I did, periodically, return to it, expanding or contracting it in various attempts to find an appropriate length for a suitable home when various opportunities (read: "call for papers") appeared. And once I encountered Professor Watts' words—his admission in *States of Violence and the Civilising Process: On Criminology and State Crime* (2016) effectively sanctioning a prolonged (period of) gestation or percolation of ideas—I permitted myself to keep "DA as CA" (as I came to refer to it) on my ongoing "To Do" list without jettisoning it to the academic scrap pile. But it would be disingenuous of me if I did not confess that as time passed, I came to wonder whether the work might have "missed its moment," so to speak—whether it no longer had anything to offer because the ideas had lost their saliency due to law enforcement suppression of (environmental) dissent and crackdown on ecodefenders, denunciation by mainstream environmental organizations of ecodefenders' tactics, and derision and scorn of ecodefenders' efforts by mainstream media experiencing increasing consolidation of ownership. At least, I felt that way, that is, until I encountered Matthew Wolfe's (2022/2023) article in *The New York Times*, where he asserts, without equivocation, that "the story of the Earth Liberation Front seems more relevant [now] than ever." As Wolfe (2022/2023) proclaims,

> After decades in which America's environmental movement confined its activities largely to rallies, marches and other lawful forms of protest, frustrated activists have begun taking a more confrontational approach. Younger groups like the Sunrise Movement and Extinction Rebellion have blockaded roads and occupied the offices of lawmakers. During the Standing Rock protests of 2016, thousands of demonstrators sought to physically impede construction of the Dakota Access Pipeline. Tim DeChristopher, a founder of the Climate Disobedience Center, which supports protesters who engage in nonviolent resistance, told me that, in the 2000s, such direct action was championed mostly by a fringe group of anarchists. (DeChristopher himself was sent to prison after placing winning

> bids at public auctions for oil and gas leases and then refusing to pay.) Now, even staid Washington-based environmental groups, sensing an increasingly unruly mood among their base, have slowly started to embrace more "radical" tactics.[2] In 2017, the Sierra Club formally lifted its 120-year ban on civil disobedience after its executive director and other senior members were arrested for strapping themselves to a gate outside the White House [internal quotation marks added].[3]

Wolfe (2022/2023) then proceeds to explain that "some climate activists have begun to openly contemplate the possibility—in their eyes, the necessity—of directly sabotaging the infrastructure of the carbon economy. Foremost among them is the academic Andreas Malm, whose recent book, 'How to Blow Up a Pipeline,' calls for smashing the tools of fossil-fuel extraction as a last-ditch means of averting ecological collapse."

Wolfe's (2022/2023) words incited in me a desire to engage with more of Malm's work. (To that point, I had really been much more familiar with his propositions in *The Progress of This Storm: Nature and Society in a Warming World* (Malm 2018) and *White Skin, Black Fuel: On the Danger of Fossil Fascism* (Malm and the Zetkin Collective 2021) and, to lesser extent, *Corona, Climate, Chronic Emergency: War Communism in the Twenty-First Century* (Malm 2020).) I reconnected with *How to Blow Up a Pipeline: Learning to Fight in a World on Fire* (2021)—the book, itself, is (or reads like) a sort of communiqué[4]—which catalyzed (perhaps accelerated is a better word choice) my return to what is now *Direct Action as Conceptual Art?: An Examination of the Role of the Communiqué for Ecodefense*.[5]

In addition to Emeritus Professor Nigel South and Professor Rob Watts, I would like to extend my thanks to Professors Michelle Brown and Eamonn Carrabine, Series Editors for Palgrave Studies in Crime, Media and Culture, for their enthusiasm for the project, as well as to Josie Taylor, Executive Editor for Criminology and Socio-Legal Studies at Palgrave Macmillan/Springer Nature, for her support and for helping this project move from paper to proposal to book.

Richmond, KY, USA Avi Brisman

Notes

1. Readers familiar with "Postscript: Drifts, Drift, and Chronicles" in my monograph, *Fieldnotes on a Study of Young People's Perceptions of Crime and Justice: Scaffolding as Structure* (2023), may notice a familiar theme here.
2. According to *The Oxford English Dictionary*, the use of the term, "radical," in the political context, can be traced to 1783, referring to:

   > [a]dvocating thorough of far-reaching political or social reform; representing or supporting an extreme section of a party; *spec.* (also with capital initial) *(a) British* belonging to, supporting, or associated with the group or movement which called for reform of the parliamentary system and other radical social and political changes in the late eighteenth and early nineteenth centuries (cf. radical reform *n.*) *(b) U.S.* belonging to a faction of the Republican Party seeking extreme action against the South during the Civil War and Reconstruction. Now more generally: revolutionary, esp. left-wing.

   Over time, the meaning of "radical" morphed, such that it came to denote not just the leftward flank of liberalism ("Radicalism" with an initial capital letter), but positions at opposite ends of political spectrums or ideologies (e.g., "radical progressives," "radical conservatives"). Thus, in modern day parlance, "radical" may be used to identify individuals with views that stand in stark contrast to each other (such as "radical left" vs. "radical right"), where the only similarity is that both are *not moderate*. This can create some confusion—especially when "radical" is used pejoratively. For example, in 2018, a chapter chair of the Sierra Club criticized United States Senator Mitch McConnell (R-KY) for using the term "radical environmentalist" to "demonize anyone who would stand up to polluters or suggest that government protections are needed to preserve our air and water for future generations" (Morris 2018: 4). Denouncing McConnell for "wrongfully disparag[ing] the good, decent, loving, moral people of the Sierra Club as 'Radical Environmentalists'," Morris (2018: 4) asserted that McConnell was the "true radical" for, *inter alia*, propping up the dying coal industry and subsidizing fossil fuel pollution—"radical *anti*-environmental policies."

   Arguably, McConnell's position is not "radical," but "extreme." As Böttícher (2017: 75) offers, "extremism" possesses "a strong palingenetic element," whereas "radicalism looks more at a golden *future for all* rather than seeking to restore an allegedly golden past [such as propping up a dying coal industry] for adherents to its creed" (emphasis added). In addition—and contra Morris (2018)—Böttícher (2017: 75) submits that "radicals," unlike "extremists," "self-define themselves as radicals." (This may, in part, stem from the fact that "while the FBI and others have increasingly used the language of 'extremism' (not 'terrorism') when discussing domestic groups, many have argued that policymakers, law enforcement, and the general public typically interpret these terms synonymously" (Loadenthal 2017: 7 (endnote omitted)); on the U.S. Department of Homeland Security's use of "extremist" to refer to "environmental activists engaged in non-violent civil disobedience targeting the oil industry . . . alongside white nationalists and mass killers," see Federman (2020).) Reflecting Bötticher's (2017) point about "radicals" willingness to self-identify as such, many academics, researchers, and scholar-activists whom I cite in this book (and who are or may be (to varying degrees) sympathetic to ecodefense as a strategy or tactic) use "radical environmentalism" and/or "radical environmentalist(s)" and/or "radical environmentalist(s)" (see, e.g., Allhoff and Buciak 2013; Amster 2006; Brown 2020, 2021; Cianchi 2013, 2015; Hays et al. 1996; Johnston and Johnston 2017, 2020; Joosse 2007, 2012, 2014, 2021; Manes 1990; Mills 1991; Pellow 2014, 2020; Salter 2011; Scarce 1990, 2006; Taylor 1998; Vanderheiden 2005; Woodhouse 2018; see also Bandow 1991; Illing 2021). Indeed, Johnston and Johnston (2017: 748n.1) offer a similar explanation for their word choice: "[they] use 'radical' because this is how activists who engage in the analyses and activities described in [their] paper are popularly referred to in the scholarly literature and media." I follow their lead and adopt their rationale, but I take the additional step of continuing to place the words "radical environmentalism" and "radical environmentalist(s)" in quotation marks. The intention is not to express disdain, irony, or skepticism of activists and their use of ecodefense to deter, reduce or stop immediately an environmental harm, risk, or threat—or to distance myself from ecodefenders—but to acknowledge that "radical environmentalism" and "radical environmentalist(s)" possess "emotive dimension[s] as well as descriptive one[s]," to quote Finlay (2009: 753), and that the term(s) is/are somewhat ambiguous, subject to contestation. The purpose of using quotation marks around the word "radical" and its accompanying noun is also to encourage readers to consider whether, given the extent of the environmental crises we are facing, the actions, strategies, and tactics discussed in the book (and the individuals or groups who undertake them) may actually not be that immoderate.

3. For a note about the Sierra Club's decision to end its 120-year-old prohibition against civil disobedience, see Burkett (2016: 9).
4. The title is a bit of a misnomer, and a number of commentators have suggested as much: "Malm's book—it's titled *How to Blow Up a Pipeline*—is obviously meant to provoke" (Illing 2021); "[Malm's] recent book, How to Blow Up a Pipeline, is a bit more nuanced than the title suggests

. . . ." (David Remnick quoted in Wickenden 2021); "Malm wants us to fight back (though I should add that there aren't any actual instructions here about how to blow anything up)" (Schlossberg 2021); and "The provocatively titled manifesto, which, to be clear, does not actually provide instructions for destroying anything, functioned both as a question—why has climate activism remained so steadfastly peaceful in the face of minimal results?—and as a call for the escalation of protest tactics like sabotage" (Marchese 2024). Essentially, while a more accurate title might have been *Why Blow Up a Pipeline* or *How to Justify Blowing Up a Pipeline*, Malm's chosen title—*How to Blow Up a Pipeline*—is certainly more compelling, more enticing. Regardless of whether Malm's book could have been titled differently, it does have the tenor of a communiqué, and I am indebted to Luke A. Fidler for encouraging me to think about Malm's *How to Blow Up a Pipeline: Learning to Fight in a World on Fire* (2021) in this way.

5. Because some of the research for this project began more than fifteen years ago, some of the online sources that I consulted became unavailable at their original uniform resource locator (URL). As a result, I attempted to find the same information at other, newer—and active (as of October 2024)—web addresses. Where I could not do so, I have indicated in the references to each chapter the last date on which I accessed the source with the now "broken" link. In many of these instances, I have retained a hard copy of the source printed from the defunct website or webpage that can no longer be found (and have noted as much with the phrase "on file with author").

References

Allhoff, Steffen W., and Sebastian K. Buciak. 2013. Action, Reaction, Escalation: Radical Environmental Organisations in the Context of Eco-terrorism and Eco-radicalism. In *Global Environmental Change: New Drivers for Resistance, Crime and Terrorism?*, Achim Maas, Balázs Bodó, Clementine Burnley, Irina Comardicea, and Roger Roffey, ed. pp. 221–242. Baden-Baden, DE: Nomos Verlagsgesellschaft.

Amster, Randall. 2006. Perspectives on Ecoterrorism: Catalysts, Conflations, and Casualties. *Contemporary Justice Review*, 9(3): 287–301. https://doi.org/10.1080/10282580600827991.

Bandow, Doug. 1991. Radical Activism Cannot Help Protect the Environment. In *The Environmental Crisis: Opposing View Points*, Neal Bernards, ed. pp. 254–260. San Diego, CA: Greenhaven.

Bötticher, Astrid. 2017. Towards Academic Consensus Definitions of Radicalism and Extremism. *Perspectives on Terrorism*, 11(3) [August]: 73–77. http://www.jstor.org/stable/26297896.

Brisman, Avi. 2023. *Fieldnotes on a Study of Young People's Perceptions of Crime and Justice: Scaffolding as Structure*. Abingdon, Oxon, UK, and New York: Routledge.

Brown, Joseph M. 2020. Notes to the Underground: Credit Claiming and Organizing in the Earth Liberation Front. *Terrorism and Political Violence*, 32(2): 237–256. https://doi.org/10.1080/09546553.2017.1364637. Published online on September 7, 2017.

Brown, Joseph M. 2021. Civil Disobedience, Sabotage, and Violence in US Environmental Activism. In *The Oxford Handbook of Comparative Environmental Politics*, Jeannie Sowers, Stacy D. VanDeveer, and Erika Weinthal, ed. pp. 356–374. Online edition, Oxford Academic, https://doi.org/10.1093/oxfordhb/9780197515037.013.34.

Burkett, Maxine. 2016. Climate Disobedience. *Duke Environmental Law & Policy Forum*, 27(1) [Fall]: 1-50. Available at: https://scholarship.law.duke.edu/delpf/vol27/iss1/1/.

Cianchi, John. 2013. *I talked to my tree and my tree talked back: radical environmental activists and their relationships with nature*. Ph.D. Thesis, School of Social Sciences, University of Tasmania (Australia).

Cianchi, John. 2015. *Radical Environmentalism: Nature, Identity and More-than-human Agency*. Basingstoke, Hampshire, UK: Palgrave Macmillan.

Finlay, Christopher J. 2009. How to do things with the word 'terrorist.' *Review of International Studies*, 35(4) [October]: 751–774. http://www.jstor.org/stable/40588072.

Hays, Scott, Michael Esler, and Carol Hays. 1996. Radical Environmentalism and Crime. In *Environmental Crime and Criminality: Theoretical and Practical Issues*, Sally M. Edwards, Terry D. Edwards, and Charles B. Fields, ed. pp. 163–181. New York and London: Garland Publishing.

Illing, Sean. 2021. The case for a more radical climate movement: Author Andreas Malm on the failures of climate activism and the need for escalation. *Vox*, October 1. Available at: https://www.vox.com/vox-conversations-podcast/22691428/vox-conversations-climate-change-andreas-malm.

Johnston, Genevieve, and Matthew S. Johnston. 2017. 'We fight for all living things': countering misconceptions about the radical animal liberation movement. *Social Movement Studies*, 16(6): 735–751. https://doi.org/10.1080/14742837.2017.1319268.

Johnston, Genevieve, and Matthew S. Johnston. 2020. 'Until every cage is empty': frames of justice in the radical animal liberation movement. *Contemporary Justice Review*, 23(4): 563–580. https://doi.org/10.1080/10282580.2020.1719361.

Joosse, Paul. 2007. Leaderless resistance and ideological inclusion: The case of the Earth Liberation Front. *Terrorism and Political Violence*, 19(3): 351–368. https://doi.org/10.1080/09546550701424042.

Joosse, Paul. 2012. Elves, environmentalism, and "eco-terror": Leaderless resistance and media coverage of the Earth Liberation Front. *Crime Media Culture*, 8(1): 75–93. https://doi.org/10.1177/1741659011433366.

Joosse, Paul. 2014. Antiglobalization and Radical Environmentalism: An Exchange on Ethical Grounds. *Ethics in Progress*, 5(1): 33–51. https://doi.org/10.14746/eip.2014.1.2. Available at https://pressto.amu.edu.pl/index.php/eip/article/view/9382/9084.

Joosse, Paul. 2021. Narratives of rebellion. *European Journal of Criminology*, 18(5): 735–754. https://doi.org/10.1177/147737081987442.

Malm, Andreas. 2018. *The Progress of This Storm: Nature and Society in a Warming World*. London and New York: Verso.

Malm, Andreas. 2020. *Corona, Climate, Chronic Emergency: War Communism in the Twenty-First Century*. London and New York: Verso.

Malm, Andreas. 2021. *How to Blow Up a Pipeline: Learning to Fight in a World on Fire*. London and New York: Verso.

Malm, Andreas, and the Zetkin Collective. 2021. *White Skin, Black Fuel: On the Danger of Fossil Fascism*. London and New York: Verso.

Manes, Christopher. 1990. *Green Rage: Radical Environmentalism and the Unmaking of Civilization*, Boston: Little, Brown and Co.

Marchese, David. 2024. How This Climate Activist Justifies Political Violence. *The New York Times*, January 16. Available at https://www.nytimes.com/interactive/2024/01/14/magazine/andreas-malm-interview.html.

Mills, Stephanie. 1991. Thoughts from the RRR. In *The Earth First! Reader: Ten Years of Radical Environmentalism*, John Davis, ed. pp. 128–132. Salt Lake City, UT: Peregrine Smith.

Morris, Tom. 2018. Chair's Corner: Radical Environmentalists. *The Cumberland*, 52(11 & 12) [November/December]: 4.

Pellow, David Naguib. 2014. *Total Liberation: The Power and Promise of Animal Rights and the Radical Earth Movement*. Minneapolis, MN: University of Minnesota Press.

Pellow, David N. 2020. Environmental justice, animal rights and total liberation From conflict and distance to points of common focus In *Routledge International Handbook of Green Criminology*, 2nd edn, Avi Brisman and Nigel South, ed. pp. 555–570. Abingdon, Oxon, UK, and New York: Routledge.

Salter, Colin. 2011. Activism as Terrorism: The Green Scare, Radical Environmentalism and Governmentality. *Anarchist Developments in Cultural Studies*, No. 1(2011): Ten Years After 9/11: An Anarchist Evaluation: 211–238. Available at: https://journals.uvic.ca/index.php/adcs/article/view/17126/7340.

Scarce, Rik. 1990. *Eco-Warriors: Understanding the Radical Environmental Movement*. Chicago, IL: Noble Press.

Scarce, Rik. 2006. *Eco-Warriors: Understanding the Radical Environmental Movement*, Updated edition. Walnut Creek, CA: Left Coast Press.

Schlossberg, Tatiana. 2021. Climate Disaster. *The New York Times Book Review*, January 24: 22. Published online as "Three Books Offer New Ways to Think About Environmental Disaster" on January 22, 2021, at https://www.nytimes.com/2021/01/22/books/review/scorched-earth-emmanuel-kreike-how-to-prepare-for-climate-change-david-pogue-how-to-blow-up-a-pipeline-andreas-malm.html.

South, Nigel, and Avi Brisman, ed. 2013. *Routledge International Handbook of Green Criminology*. Abingdon, Oxon, UK, and New York: Routledge.

Taylor, Bron. 1998. Religion, violence, and radical environmentalism: From earth first! to the Unabomber to the earth liberation front. *Terrorism and Political Violence*, 10(4): 1–42. https://doi.org/10.1080/09546559808427480.

Vanderheiden, Steve. 2005. Eco-terrorism or Justified Resistance? Radical Environmentalism and the "War on Terror." *Politics & Society*, 33(3): 425–447. https://doi.org/10.1177/0032329205278462.

Watts, Rob. 2016. *States of Violence and the Civilising Process: On Criminology and State Crime*. London: Palgrave Macmillan/Springer Nature.

Watts, Rob. 2020. *Criminalizing Dissent: The Liberal State and the Problem of Legitimacy*. Abingdon, Oxon, UK, and New York: Routledge.

Wickenden, Dorothy. 2021. Andreas Malm on the Environmental Movement and "Intelligent Sabotage." *The New Yorker*, September 27. Available at: https://www.newyorker.com/podcast/political-scene/andreas-malm-on-the-environmental-movement-and-intelligent-sabotage.

Wolfe, Matthew. 2022/2023. Backfire. *The New York Times Magazine*, May 29: 38. Published online as "The Rise and Fall of America's Environmentalist Underground" on May 26, 2022 [updated June 15, 2023], at https://www.nytimes.com/2022/05/26/magazine/earth-liberation-front-joseph-mahmoud-dibee.html.

Woodhouse, Keith Makato. 2018. *The Ecocentrists: A History of Radical Environmentalism*. New York: Columbia University Press.

A Note About Endnotes

Law reviews—student-edited periodicals—provide academic and practicing legal professionals with the opportunity to analyze, criticize, and otherwise comment on recent "court decisions, unresolved issues of law, and other topics of interest to the legal community" (Swygert and Bruce 1985: 739).[1] Much has been written extolling law reviews' pedagogical benefits and influence on the development of law (Swygert and Bruce 1985: 739–740, 787–790; see also Cramton 1986): the right to privacy, for example, originated in an article by Samuel D. Warren and Louis D. Brandeis in the *Harvard Law Review* in 1890.[2] On the other hand, much has also been written dismissing their impact (Liptak 2007) and criticizing their content, quality, accessibility, student-run qualities (evaluation, revision, and publication), and style (see, e.g., Austin 1987: 1132; Rodell 1937, 1962; Swygert and Bruce 1985: 740n.7; cf. Zenoff 1986).

Within this body of literature, there exists an "enjoyable thread" (Hutchens 1998: 1093n.17) of law review articles ruminating on a particular aspect of law review style: the use—some (might) say "misuse" or "abuse" (Austin 1990: 1011)—of citations and footnotes. As McPherson (2019: 457n.1) admits, "our footnotes, at once detail-oriented, comprehensive, and accurate, ofttimes go overboard in our attempts to impress one another—and to see our work get published. Those of us who seek to share with the world our wealth of legal knowledge learn quite early in our careers that the secret to publication is often footnotes, footnotes, and more footnotes" (citing Shapiro and Pearse 2010: 1485).[3]

The late Honorable Abner J. Mikva, Chief Judge of the United States Court of Appeals for the District of Columbia, likened the use of footnotes in legal writing to a "fungus [that] has magnified all of the short-comings of legal writing" (Mikva 1985: 647). He regarded the use of footnotes in judicial opinions as an "abomination" and posited that "[i]f footnotes were a rational form of communication, Darwinian selection would have resulted in the eyes being set vertically rather than on an inefficient horizontal plane" (Mikva 1985: 647, 648).

Mikva's (1985) assertion subsequently led C. Steven Bradford (1994: 24), the Henry M. Grether, Jr., Professor of Law Emeritus at the University of Nebraska-Lincoln College of Law, to retort:

> As every law review editor knows, footnotes are the most important part of the article. An article without footnotes is like a footnoteless article. Don't make what is known in the legal community as the "Mikva Mistake." Judge Abner J. Mikva, a noted federal jurist, once wrote an article without footnotes and, as a result, never made it to the Supreme Court. If you want to be successful, good footnotes are a must.
>
> The substantive content of your footnotes is irrelevant. What's important is the style.[4]

Other academic disciplines rarely display law's affinity—some might say *infatuation*—with footnotes. Historian G.W. Bowersock (1984: 54), Professor Emeritus at the Institute for Advanced Study at Princeton, contemplates the "art of sophisticated annotation" in a variety of literary milieus in "The Art of the Footnote": accepting its use when its contents do "not fit seamlessly into a fabric of sentences" (1984: 55), yet cautioning against numerous and long entries, as well as ones using *supra*, *infra*, *ibid*. and *id*. (1984: 59, 55);[5] proclaiming it "a work of art and an instrument of power . . . [i]n the hands of a master" (1984: 54 (referring to eighteenth-century English Edward Gibbon)), but agreeing with actor, playwright, and composer Noel Coward's aphorism that "[e]ncountering [a footnote] is like going downstairs to answer the doorbell while making love."[6] The late Herma Hill Kay (1990: 419), the Barbara Nachtrieb Armstrong Professor of Law at the University of California, Berkeley School of Law, generalized that "[s]ocial scientists are notoriously down on law review style." Similarly, Laura Nader (1979: 998n.†), Professor of Anthropology at the University of California, Berkeley, has commented that "in many respects, law review form is not conducive to the expression and development of ideas by persons trained in other disciplines"—a point she makes in a law review *footnote*, no less.

Direct Action as Conceptual Art?: An Examination of the Role of the Communiqué for Ecodefense is not a law review article—and, in fact, it is actually shorter than some of the law review articles that I wrote earlier in my career. It also does not contain a single footnote. But I have inserted a large number of *endnotes* of varying lengths. The goal has not been to inspire today's Noel Coward or John Barrymore to compare encountering an endnote to running out the downstairs master bedroom to answer a knock at the *backdoor* while in the middle of sex.[7] Nor has it been to try to earn the sobriquet of the "David Foster Wallace of Criminology," as a senior editor of a prominent publisher once suggested upon reading my sample chapters of a proposed manuscript I had submitted to him.[8] Rather, I wanted to write a book whose central claims could be gleaned in one sitting (in the main text), while also pointing readers to more ideas, information, sources, and support for those interested in particular points or propositions (to be found in the endnotes). Thus, the principal arguments of *Direct Action as Conceptual Art?: An Examination of the Role of the Communiqué for Ecodefense* appear in the body text; additional comments, examples, and pensées at the end of each chapter.

Notes

1. As Swygert and Bruce (1985: 739) explain—and it will soon become clear why I might have chosen to indicate this in a note—"[m]ost accredited law schools in the United States publish a student-edited law review . . . ." In a footnote at the end of their sentence, Swygert and Bruce (1985: 739 n.3) make clear, in case there is any confusion, that "[t]he term 'law review' is used in [their] article to refer to student-edited scholarly reviews, journals, quarterlies, or similar periodicals published by law schools. The typical law review is divided into two major sections. The first section contains articles composed by experienced professionals and edited by law students. The second is comprised of works written and edited by law students" (citing Fidler 1983; Lee 1956).
2. It is one of the most-cited law review articles of all time and is "commonly regarded as the most influential of all law review articles, virtually originating the tort of invasion of privacy" (Shapiro and Pearse 2012: 1489, 1503).
3. McPherson's (2019) confession appears in a footnote, of course. Her article—its "Introduction, Body, and Conclusion"—consists of one

sentence: "Law Review articles have too many footnotes," followed by a single footnote that spans three pages and contains almost 2000 words. McPherson's article is not, however, the shortest law review article ever written. Jensen's (2000: 156) article, entitled "The Shortest Article in Law Review History," contains three words: "This is it" (followed by two footnotes). Morris's (2000: 310) article, "The Shortest Article in Law Review History: A Brief Response to Professor Jensen," is comprised of only two words: "Not so!" (followed by one footnote).

4. Other key contributions to the scholarly analysis of footnotes within the field of "citology" or "legal citology" (see Balkin and Levinson 1996) include "The Footnote," by J.M. Balkin (1989), "Bottomheavy: Legal Footnotes," by Joan Ames Magat (2010), and "The Under-Theorized Asterisk Footnote," by Charles A. Sullivan (2005).
5. Law review publications make frequent use of *supra*, *infra*, and *id.*, but generally not *ibid*. Cases, however, sometimes use *ibid*. (see, e.g., *Roper v. Simmons*, 543 U.S. 551, 596 n.* (2005) (O'Connor, J., dissenting) (using the signal *ibid*.)).
6. Others have made similar observations. For example, John Barrymore (1882-1942), the American actor of stage, screen, and radio, once observed that reading footnotes is akin to "having to run downstairs to answer the doorbell during the first night of the honeymoon" (quoted in Brisman 2016: vii).
7. Indeed, as Jonathan Russell Clark (2015) points out rather astutely, "[t]wo of the most famous novels to use footnotes actually contain no footnotes at all. What Vladimir Nabokov's *Pale Fire* and David Foster Wallace's *Infinite Jest* have are *endnotes*. The distinction may seem facile, but endnotes leave the text clean and present the reader with a different kind of choice as well as a very distinctive reading experience."
8. I took this senior editor's comment as a compliment—and a sign of encouragement. It was not intended as such. Three of the five reviews that the senior editor received and shared with me decried my "over-reliance on referencing and footnoting" and found them "distracting." Reviewer #1: "a good third of the text were footnotes and the over use of such lengthy quotations in his chapter would warrant criticism in a UG text." Reviewer #3: "combining in-text referencing and use of footnotes is a wee bit superfluous." Reviewer #4: "I have a slight reservation about the volume of footnotes in the sample chapters—I tend to think that these should be kept to a minimum. If the information contained in many of them is important it should be worked in the main text—if not—they should be deleted."

   I made promises to alter the style. To Reviewer #1: "I will . . . trim the footnotes." To Reviewer #3: "I will revise . . . with an eye toward pruning in-text references and footnotes." To Reviewer #4: "I intend to prune the footnotes and references."

Two months later, I received a contract.

By the time I submitted the final manuscript, it had undergone substantial changes—but not to the notes.

For readers interested in comments and reflections on David Foster Wallace's use of footnotes, see, e.g., Hesse (2008); Staff Writer (2007). Of course, one could (also) (just) read *Infinite Jest* (1996) or his short stories,"The Depressed Person" and "Octet," in *Brief Interviews with Hideous Men* ([1999] 2007).

References

Austin, Arthur D. 1987. Footnotes as Product Differentiation. *Vanderbilt Law Review*, 40(5) [October]: 1131–1155. Available at: https://scholarship.law.vanderbilt.edu/vlr/vol40/iss5/3.

Austin, Arthur. 1990. Footnote Skulduggery and Other Bad Habits. *University of Miami Law Review*, 44(4): 1009–1031. Available at: https://repository.law.miami.edu/umlr/vol44/iss4/3.

Balkin, J.M. 1989. The Footnote. *Northwestern University Law Review*, 83(1&2) [Fall 1988/Winter 1989]: 275–320. http://hdl.handle.net/20.500.13051/2227; https://openyls.law.yale.edu/handle/20.500.13051/2227.

Balkin, J.M., and Sanford Levinson. 1996. How to Win Cites and Influence People. *Chicago-Kent Law Review*, 71(3) [April] [Symposium on Trends in Legal Citations and Scholarship]: 843–869. https://scholarship.kentlaw.iit.edu/cklawreview/vol71/iss3/10.

Bowersock, G. W. 1984. The Art of the Footnote. *The American Scholar*, 53(1) [Winter]: 54–62. http://www.jstor.org/stable/41211579.

Brisman, Avi. 2016. *Geometries of Crime: How Young People Perceive Crime and Justice*. London: Palgrave Macmillan.

Clark, Jonathan Russell. 2015. On the Fine Art of the Footnote: From Nabokov to Danielewski, Beyond the Experimental. *Literary Hub*, June 3. Available at: https://lithub.com/the-fine-art-of-the-footnote/.

Cramton, Roger C. 1986. "The Most Remarkable Institution": The American Law Review. *Journal of Legal Education*, 36(1) [March]: 1–10. https://www.jstor.org/stable/42892822.

Bradford, C. Steven. 1994. As I Lay Writing: How to Write Law Review Articles for Fun and Profit: A Law-and-Economics, Critical, Hermeneutical, Policy Approach and Lots of Other Stuff That Thousands of Readers Will Find Really Interesting and Therefore You Ought to Publish in Your Prestigious, Top-Ten, Totally Excellent Law Review: [this space reserved]. *Journal of Legal Education*, 44(1) [March]: 13–34. Available at: https://digitalcommons.unl.edu/lawfacpub/87; https://digitalcommons.unl.edu/cgi/viewcontent.cgi?referer=&httpsredir=1&article=1086&context=lawfacpub.

Fidler, Josh E. 1983. Law-Review Operations and Management: AN EMPRICAL STUDY OF THE NEW YORK UNIVERSITY LAW REVIEW ALUMNI ASSOCIATION. *Journal of Legal Education*, 33(1) [March]: 48–63. https://www.jstor.org/stable/42897833.

Hutchens, Walter C. 1998. Alienage Jurisdiction and the Problem of Stateless Corporations: What is a Foreign State for Purposes of 28 U.S.C. § 1332(A)(2)? *Washington University Law Quarterly*, 76(3): 1067–1093. Available at: https://openscholarship.wustl.edu/law_lawreview/vol76/iss3/5/; https://journals.library.wustl.edu/lawreview/article/6909/galley/23742/view/.

Hesse, Monica. 2008. Foster Wallace: beyond a footnote. *The Seattle Times*, September 16. Available at: https://www.seattletimes.com/entertainment/books/appreciation-foster-wallace-beyond-a-footnote/.

Jensen, Erik M. 2000. The Shortest Article in Law Review History. *Journal of Legal Education*, 50(1) [March]: 156. https://www.jstor.org/stable/42893636.

Kay, Herman Hill. 1990. In Defense of Footnotes. *Arizona Law Review*, 32(3): 419–429. Available at: https://lawcat.berkeley.edu/record/1113828?ln=en&v=pdf.

Lee, Richard H. 1956. ADMINISTRATION OF THE LAW REVIEW. *Journal of Legal Education*, 9(2): 223–230. https://www.jstor.org/stable/42891024.

Liptak, Adam. 2007. When Rendering Decisions, Judges Are Finding Law Reviews Irrelevant. *The New York Times*, March 19. Available at: https://www.nytimes.com/2007/03/19/us/19bar.html?searchResultPosition=1.

Magat, Joan Ames. 2010. Bottomheavy: Legal Footnotes. *Journal of Legal Education*, 60(1) [August]: 65–105. https://www.jstor.org/stable/42894154.

McPherson, Lori. 2019. Law Review Articles Have Too Many Footnotes. *Journal of Legal Education*, 68(2) [Winter]: 457–459. https://www.jstor.org/stable/26891022.

Mikva, Abner J. 1985. Goodbye to Footnotes. *University of Colorado Law Review*, 56 [Summer]: 647–653.

Morris, Grant H. 2000. The Shortest Article in Law Review History: A Brief Response to Professor Jensen. *Journal of Legal Education*, 50(2) [June]: 310. https://www.jstor.org/stable/42893645.

Nader, Laura. 1979. Disputing Without the Force of Law. *Yale Law Journal*, 88(5) [April]: 998–1021. Available at: https://openyls.law.yale.edu/bitstream/handle/20.500.13051/15890/48_88YaleLJ998_April1979_.pdf?sequence=2&isAllowed=y; https://escholarship.org/uc/item/14j3c4w2.

Rodell, Fred. 1937. Goodbye to Footnotes. *Virginia Law Review*, 23(1) [November]: 38–45. https://doi.org/10.2307/1067669; http://www.jstor.com/stable/1067669.

Rodell, Fred. 1962. Goodbye to Law Reviews. Revisited. *Virginia Law Review*, 48(2) [March]: 279–290. https://doi.org/10.2307/1071179; https://www.jstor.org/stable/1071179.

Shapiro, Fred R., and Michelle Pearse. 2010. The Most-Cited Law Review Articles of All Time. *Michigan Law Review*, 110(8) [June]: 1483–1520. Available at: https://repository.law.umich.edu/mlr/vol110/iss8/2.

Staff Writer. 2007. The footnotes of "Infinite Jest." *Sarasota Herald-Tribune*, August 30. Available at: https://www.heraldtribune.com/story/news/2007/08/30/the-footnotes-of-infinite-jest/28642133007/.

Sullivan, Charles A. 2005. The Under-Theorized Asterisk Footnote. *Georgetown Law Journal*, 93(3) [March]: 1093–1116.

Swygert, Michael I., and Jon W. Bruce. 1985. The Historical Origins, Founding, and Early Development of Student-Edited Law Reviews. *Hastings Law Journal*, 36 [May]: 739–791.

Warren, Samuel D., and Louis D. Brandeis. 1890. The Right to Privacy. *Harvard Law Review*, 4(5) [December 15]: 193–220. https://doi.org/10.2307/1321160; https://www.jstor.org/stable/1321160.

Wallace, David Foster. 1996. *Infinite Jest*. New York: Little, Brown and Company.

Wallace, David Foster. [1999] 2007. The Depressed Person. In *Brief Interviews with Hideous Men*, pp. 37–69. New York: Back Bay Books/Little, Brown and Company.

Wallace, David Foster. [1999] 2007. Octet. In *Brief Interviews with Hideous Men*, pp. 131–160. New York: Back Bay Books/Little, Brown and Company.

Zenoff, Elyce H. 1986. I Have Seen the Enemy and They Are Us. *Journal of Legal Education*, 36(1) [March]: 21–23. https://www.jstor.org/stable/42892826.

Contents

CHAPTER 1

Introduction

The following appeared in a March 2008 newspaper: "Five unoccupied houses near Maltby, Wash., were destroyed or severely damaged Monday by fire." Destructive fire often makes local or regional news (see, e.g., Leach 2023; Parker and Edwards 2024). Sometimes, destructive fire can make national or international headlines—especially in the case of deadly, fast-moving wildfires (see, e.g., Archibold 2009; Jolly et al. 2024; Karlamangla et al. 2024; Nicas and Fuller 2018; Nostrant 2024)[1] or in instances where the fire is associated with previously peaceful political protest (see, e.g., Associated Press 2014). But news about the March 2008 fire in Maltby, Washington, appeared in what was then known as the "Week in Review" section of *The New York Times*—a section of the Sunday paper usually devoted to summarizing and discussing the *major* stories from the week.[2]

Why did this story stand out from the hundreds of other stories from the week? Why did this fire receive national attention instead of any of the dozens of other fires that likely occurred during the same time frame?

Even if we consider that the full line actually read, "Five unoccupied houses near Maltby, Wash., were destroyed or severely damaged Monday by fires that officials said were deliberately set," it is not readily apparent why this story attained such a prominent place (page 2) in such a prominent section of such a prominent newspaper.[3] While theories about news

A. Brisman, *Direct Action as Conceptual Art?*, Palgrave Studies in Crime, Media and Culture,
https://doi.org/10.1007/978-3-031-76634-3_1

selection criteria might (begin to) provide some clues (see, e.g., Chibnall 1977; Cohen and Young 1973; Gans 1979; Graber 1980; Ross 2007; see also Barak 1988, 1994, 1996, 2007, 2017; Ferrell 1998; Greek and Henry 1997; Hayward and Ilan 2017; Henry 1994; Kudlac 2007), even with such tools, the fact that there were no casualties or injuries and that the fires occurred in a town with a population that, at the time, contained fewer than 10,000 residents[4] might leave the most experienced journalist, editor, or social scientist to conclude that deadlines, time constraints, and layout led to the decision (see Gans 1979: 146).

The paragraph that followed the opening line somewhat diminishes the suspense:

> The police and F.B.I. investigators said the fires might have been ecoterrorism: A handmade sign at the scene indicated that a nebulous group called the Earth Liberation Front was responsible, deriding the builder's claims that the houses had been built in a 'green,' that is, environmentally sensitive, way and calling them McMansions. Local residents also opposed the luxury development in a wooded wetland; there had been earlier cases of vandalism; and the houses had sat on the market unsold for some time.

There have been instances in which law enforcement authorities and the media have erroneously attributed arson to the Earth Liberation Front (the "E.L.F." or "ELF")[5]—an international, underground movement consisting of anonymous, autonomous cells who carry out clandestine direct action (usually in the form of economic sabotage) in defense of the planet (North American Earth Liberation Front Press Office [NAELFPO], n.d., *Frequently Asked Questions About the Earth Liberation Front*; NAELFPO, n.d., *What is the Earth Liberation Front?*; see also Brisman 2008: 755–756; Brown 2020: 237, 242, 249, 250, 2021: 365; Joosse 2012: 76–79; 2014: 34; 2021: 739; Manson 2024: 3; Pickering 2007: 43; Vanderheiden 2005: 445; see generally Johnston and Johnston 2017: 738; 2020: 564, 567, 576; cf. Resistance Magazine 2009a: 9).[6] But with the Maltby fires, the E.L.F. took credit for this action and explained its rationale (albeit briefly). The sign to which the "Week in Review" summary made reference—a red spray-painted bedsheet found hanging on a fence at the site—declared:

Built Green?
Nope BLACK!
McMansions + R.C.D.'s
r not GREEN
ELF[7]

Thus, we have a situation in which a nondescript fire in an unremarkable town generated national attention because of the alleged intentions and philosophical orientations of the actors.[8] In other words, had rebellious teenagers set the fires out of boredom, it is likely that few outside of Maltby, Washington, would have learned of the conflagration. But because an entity that is considered a "domestic terrorist organization"[9] was credited with the action, the Maltby fires became national news, the E.L.F. gained some exposure, and the public was alerted to the negative environmental impact of development and sprawl.

What should we make of the Maltby fires? Putting aside the fact that the fires gained national exposure, should we treat them the same as any other instance of malicious burning of residential, commercial, or industrial property in which there were no injuries or deaths but significant destruction occurred? Should we—and, if so, how should we—distinguish between the Maltby fires, fires committed by anomic teenagers, and fires committed out of anger or a desire for revenge?[10] Does the E.L.F.'s designation as a "domestic terrorist organization"—indeed, as the "number one" domestic terrorism/t threat[11]—somehow make the fires qualitatively worse? Is there something justifiable about these "crimes in the name of the environment" (Brisman 2008: 759; Hays et al. 1996: 163) because the actors felt that far more serious harm was occurring to the wooded wetland as a result of the development? Finally, should we—and, if so, *why* should we—revisit these fires in early 2025—more than fifteen years after they occurred?

In previous work, I drew on art criticism, art history, and aesthetics to explore an environmental issue. For example, in "The Aesthetics of Wind Energy Systems" (Brisman 2005a), I considered opposition to wind farms on visual or aesthetic grounds, argued that such antagonism is inconsistent with the aesthetic sensibility we bring to our appreciation of fine art, and suggested that we look to art (particularly, seventeenth-century Dutch painting and the grand-scale, site-specific, environmental art projects of the Bulgarian artist Christo (Christo Vladimirov Javacheff (1935–2020)) to guide us in our aesthetic assessment of wind turbines.

Similarly, in "'Green Harms' as Art Crime, Art Criticism as Environmental Dissent" (Brisman 2011a), I examined responses to *The Physical Impossibility of Death in the Mind of Someone Living*—a piece by the contemporary British artist, Damien Hirst (1965–), consisting of a 14-ft tiger shark (caught by a fisherman commissioned to do so) immersed in formaldehyde in a vitrine of glass and steel—to demonstrate how a critique of it on artistic grounds (and how art criticism, more generally) might be employed as an effective tool for dissent, especially in cases involving art that causes nonhuman animal suffering and/or ecological or environmental harms.[12]

Direct Action as Conceptual Art?: An Examination of the Role of the Communiqué for Ecodefense follows the line of inquiry of those prior pieces and continues a broader interest in developing green cultural criminology, which attempts to: (1) assign greater consideration to the way(s) in which environmental crimes, harms and disasters are constructed and represented by news media (e.g., online, print), social media, and in popular cultural forms (e.g., art, books, films); (2) dedicate increased attention to patterns of consumption, constructed consumerism, commodification of nature, and related market processes; and (3) devote heightened concern to the contestation of space, transgression, and resistance, in order to analyze the ways in which environmental harms are opposed in/on the streets and in day-to-day life.[13] This book also draws inspiration from recent work on the criminalization and suppression of activism, dissent, and protest,[14] as well as visual criminology, which attends to the relationship of power to representations and images of crime and crime control (Brown and Carrabine 2017a, 2019; see also Brown 2014; Carrabine 2012; Lundberg 2021).[15]

In *Direct Action as Conceptual Art?: An Examination of the Role of the Communiqué for Ecodefense*, I investigate the meaning of ecodefense (understood here to include direct action in defense of or on behalf of the environment and nonhuman animals (e.g., economic sabotage, property destruction, lab raids, "live liberations")) by examining the communiqué—the written statement or press release often issued after the action, and usually longer and more detailed than the sign that appeared at the fires in Maltby, Washington. To further comprehend the relationship of the act of ecodefense to the communiqué, I compare the communiqué to the use of written statements by Conceptual artists of the 1960s and 1970s—anti-formalist artists who emphatically rejected aesthetics and sought to promote ideas about art over art objects (see Morgan 1994: xiii,

1, 33; 1996: 2).[16] Thus, this book sets forth and then seeks to evaluate the following analogy:

> **act of ecodefense: communiqué:: work of conceptual art: written declaration/statement**

Direct Action as Conceptual Art?: An Examination of the Role of the Communiqué for Ecodefense is divided into two main parts. Chapter 2, "Ecodefense and the Action-Communiqué Relationship," opens with a definitional discussion, noting the way that different theorists (e.g., Brown 2021; Farrer and Klein 2022; Johnston and Johnston 2017; Joosse 2014; Manson 2024; Martin 1990; McIntyre 2021; Smith 2018; Vanderheiden 2005) have described and delineated between concepts and terms such as "civil disobedience," "conscientious wrongdoing," "direct action," "economic sabotage" ("eco-sabotage" or "ecosabotage" or "ecotage"), "environmental sabotage," "environmental blocking," "forceful or violent environmental sabotage," "monkeywrenching," "nonviolent direct action," and "obstructive direct action." From here, the chapter offers a brief overview of different types of ecodefense before considering four (4) specific actions and the communiqués that accompanied them. After comparing and assessing these four (4) communiqués, this chapter then endeavors to extract some features of the action-communiqué relationship and ecodefense, more generally. Chapter 3, "Conceptual Art: The Manifestation of an Idea and the Documentation of Art, an Event or (a) Process," attempts to compare the action-communiqué relationship to the use of written declarations and documents by Conceptual artists. This chapter begins with specific examples of Conceptual art before taking a step back and summarizing key features of the Conceptual art phenomena. It then attempts to suggest how a consideration of Conceptual art can illuminate features of ecodefense and to propose why such an undertaking is useful to our understanding of direct action in the name of and on behalf of the environment. In Chapter 4, "Conclusion," I summarize the main argument of the book and contemplate why the communiqué may take on heightened significance as frustration continues to mount with the failures of the political process in the worsening climate crisis.

Notes

1. In contrast to a "beneficial fire" (Hankin et al. 2023; Jordan 2023; Selig 2024) or "controlled burn," such as the one depicted on the cover of this book, "wildfire" refers to unplanned/unwanted, uncontrolled, and unpredictable fires in natural areas. Technically, "wildfire" is the umbrella term for such uncontrolled fires, with "bushfire" and "forest fire" used to identify the specific kind of "wildfire" (based on the type of vegetation present). In practice, however, the terms are often employed interchangeably. Because such fires (regardless of the label) are often unanticipated, difficult to contain, and can spread quickly—and, thus, can cause damage to built environments and destruction of ecosystems, including loss of human and nonhuman life (see, e.g., Gergis 2024; Hagmann et al. 2021; Wu et al. 2023; see generally Hessburg et al. 2021; Prichard et al. 2021)—they frequently attract significant media attention. For example, Archibold's article appeared on page A10 of *The New York Times* on September 1, 2009. A color picture of the wildfire north of Los Angeles with the caption, "Wildfire Grows in Southern California," graced the front page of that day's paper.

   Writing about the role of "the visual" in constructing social perceptions of the environment and environmental change, crime and harm, McClanahan (2021: 58–59) describes how during the devastating 2019–2020 Australian bushfire season—Australia is particularly prone to bushfires due to its hot, dry climate—"images of the fires were everywhere in the media landscape; heart-stopping pictures of crowds of fleeing people gathered on beaches, of . . . landscape[s] transformed into . . . smoldering hellscape[s], of charismatic fauna like kangaroos and koala bears fleeing or perishing, or of fatigued firefighters working against the impossible conditions of the burn, which ultimately consumed nearly 50 million acres of the continent (Tin et al. 2021)." (For more on the catastrophic Australian bushfire season of 2019–2020, dubbed the "Black Summer," see, e.g., Chester (2020); Cousins (2020); Evershed and Ball (2019); Hope (2020); *The Lancet* (2020); Morton (2019, 2020); Nolan et al. (2020); Parsons and Goldman (2020); Read and Denniss (2020); Vardoulakis et al. (2020); Wallace-Wells (2024); Walter et al. (2020); for a brief criminological discussion, see Armstrong et al. (2020); for a more in-depth criminological analysis, see Lundberg (2022, 2023).)

As global warming has caused a decrease in rainfall and an increase in extremely hot temperatures (at least in some parts of the world, such as Australia), wildfires (bushfires) have become—and will continue to become—(even) more intense and more destructive (see, e.g., BBC News 2019; Langellier 2024; Nolan et al. 2020; Parsons and Goldman 2020; Paddison 2019/2020; Wallace-Wells 2024). Accordingly, newspapers and other media sources have reported—and likely will continue to do so—on more than one dramatic fire *at the same time*. For example, in late July 2024, *The New York Times* published articles on both a wildfire in Northern California (Knight et al. 2024) and one in Jasper, Alberta, Canada (Austen and Isai 2024). In September 2024, Jolly, Selig, and Parsons reported on three (3) major wildfires occurring in Southern California at the time—the "Bridge fire," the "Line fire," and the "Airport fire." The next day, Karlamangla, Jolly, and Cowan (2024) put these fires in an even larger context, noting that "more than 65 large fires were blazing across the United States on Wednesday [September 11, 2024], mostly in the West," and that "[n]o state is battling more large fires than Oregon, where there are 24 burning. A close second is Idaho, with 22."

2. The section is now known as the "Sunday Review." Although the format has changed slightly, the scope and purpose of the new "Sunday Review" section is more or less the same as the old "Week in Review."
3. The headline at the top of the page read "Murky Mischief | March 2–8." Five images were printed underneath it, each with its own heading (from left-to-right): "ARSON," "EXPLOSION," "POISON," and "CONTAMINATION." The line quoted above in the text appeared under the image for "ARSON."
4. Based on 2000 United States Census data, Maltby, Washington, had a population of 8,472; 2010 United States Census data indicate a population of 10,830, and 2020 United States Census data show a population of 11,277 (U.S. Census Bureau 2000, 2010, 2020).
5. For example, when fires damaged more than forty unoccupied houses (and completely destroyed twelve unoccupied houses) in a

development that had been criticized by environmentalists because of its proximity to a nature preserve, investigators initially suspected that the fires had been set by "environmental extremists" (Gately 2004a). In fact, the fires were set by a group of racist young men who were seeking fame for their fledgling gang and who were angry at the developers for past slights and disdainful of the African-American homeowners who were to move into the area (see, e.g., "A Quest for Fame Is Said to Lead to Arsons" 2005; Associated Press 2004a, b, 2005b; Barringer 2004a, b; Gately 2004a, b, c, d, 2005a, b, c, d, e, 2006a, b; Lyman and Gately 2005; see generally Associated Press 2005a). Even after the motives of the arsonists were determined, the media continued to misrepresent the fires as the work of "environmental radicals." For example, Wall's (2005) article, "Burning Down the House," in the Atlanta, Georgia, monthly paper, *Creative Loafing*, attributed the December 2004 fires in Maryland to "extreme environmentalists"—an error that went uncorrected until my letter to the editor a couple of weeks later (Brisman 2005b).

6. For a general discussion of cellular groups, leaderless resistance, and "flat" organizational forms, see Joosse (2007, 2012, 2014, 2021); see also Brown (2020: 239–240).
7. "R.C.D.'s" refer to "rural cluster developments" (Joosse 2014: 39; Yardley 2008), and "advocates say help prevent sprawl by limiting development density in rural areas" (Yardley 2008).
8. I write "alleged" because there have been instances in which "the E.L.F." claimed responsibility for an action only to have its claims questioned by others suggesting that the actions could have been undertaken to discredit the movement, generate sympathy for the victims/property owners, and receive insurance money (see, e.g., Glick 2003). To my knowledge, no one suggested that the fires in Maltby, Washington, were actually set by the developer seeking insurance money—although the homes were unoccupied at the time and had been on the market for over a year.
9. Although the E.L.F. has been around since the early 1990s (formed in Brighton, England, as an offshoot of the Earth First! movement) (see, e.g., Joosse 2012: 78; 2014: 38; Loadenthal 2017: 2; Pickering 2007: 42–43; see also Brown 2020: 241, 2021: 365; Vanderheiden 2005: 426; see generally Woodhouse 2018: 228), it

was not until January 2001 that the Federal Bureau of Investigation ("F.B.I.") upgraded the E.L.F. to a "terrorist organization" (see Combs 2006: 180). And while it bears mention that individuals involved with Earth First!, the progenitor of the E.L.F., were considered "terrorists" well before the F.B.I. designation (see, e.g., Badolato 1991), Wolfe (2022/2023) reminds us that "[f]or years, the phrase 'eco-terrorism' had more often been used to describe violence *against* the natural world than vandalism committed *in its name*" (emphasis added). As Wolfe (2022/2023) explains, "[w]hen Saddam Hussein dumped millions of gallons of oil into the Persian Gulf, President George H.W. Bush denounced it as an act of 'environmental terrorism.' Yet, by the mid-90s, conservatives had begun warning of the eco-terrorism perpetrated by radical environmentalists. News outlets, including The New York Times"—the very publication in which Wolfe's exposé appears—"regularly described the ELF as terrorists—despite the fact, as activists pointed out, that the ELF had never killed anyone."

In this book, as in previous work, I refrain from using the word "terrorism" to refer to actions taken in the name of and for the sake of protecting the Earth, as well "terrorist" as a label for those who undertake such actions. As I have argued, "[u]sing the word *terrorism* to refer to actions taken in the name of the Earth and for the sake of environmental protection—something that is usually done by government officials and corporate officers—is unfortunate and exceedingly and, arguably, intentionally misleading: it conflates actions taken to thwart environmental destruction—actions that have never resulted in human injury or death—with those acts that use the environment as a tool for indiscriminate violence or threatened violence to large numbers of innocent civilians for the purpose of causing disruption, panic, harm and death" (Brisman 2008: 757 (emphasis in original; endnotes omitted)).

Relatedly—and rather oddly—the U.S. Environmental Protection Agency's ("E.P.A.") "most-wanted list"—"E.P.A. Fugitives" (https://www.epa.gov/enforcement/epa-fugitives)—names people who have been charged with "assaulting nature" (*The San Diego Union-Tribune* 2008/2016), but does not include anyone involved with a domestic "terrorist" organization. (In 2005, the F.B.I. "deemed the animal liberation movement to be America's No. 1 domestic terrorism threat" (Lennard 2019;

see also Brown 2019).) In fact, the "Current EPA Fugitives"—"[d]efendants charged with environmental crimes or violations of the U.S. Federal Criminal Code [who] flee the court's jurisdiction and/or the USA rather than face prosecution or to serve a sentence" (https://www.epa.gov/enforcement/epa-fugitives)—contains the profiles of just seventeen (17) "fugitives" (as of April 2025). This begs the question of whether those who have undertaken actions "in the name of and for the sake of protecting the Earth" (and who have not been apprehended) are not, in fact, "terrorists" (because one might assume that a "terrorist" would feature on a "most-wanted list"). Of course, the anemic list also sends the widely inaccurate message that environmental crimes have not been committed very frequently (and, rather perversely, that we are not experiencing a "climate crisis" (Harvey 2023; Illing 2021; Malm 2021: 21, 38, 57, 76, 87, 96, 106, 122, 142, 153; Marchese 2024; Moore 2020; see also Burkett 2016: 2) or that the E.P.A. is incredibly effective at enforcing environmental laws—so much so that there remains but a score of offenders on the loose—or, sadly (and more realistically), that adequate resources have not been devoted to E.P.A. criminal enforcement. Were the E.P.A. genuinely interested in generating a "most-wanted list" of environmental criminals, the agency could start with the chief executive officers, chief financial officers, and chief operations officers of major fossil fuel companies.

For a note about the origins of the "eco-terrorist" or "ecoterrorist" label, see, e.g., Amster (2006); Salter (2011: 212n.1); Smith (2008: 545–546). For a discussion of the difficulties in—and problems with—defining "terrorism," more broadly, see Finlay (2009 [especially p. 752 n. 7]); see also Chapter 4, n. 2 and accompanying text; Clark and Raustiala (2007); Curra (2017: 96–108, 165–166); Furedi (2009); Gleick (2006); Jämte and Ellefsen (2020: 397, 400); Loadenthal (2017: 2–6); Mahler (2008); McLaughlin (2013); Ruggiero (2017: 313–314); Salter (2011: 212, 218, 219, 226, 227, 228, 229, 230, 232); Sciolino and Schmitt (2008); Selmini and Di Ronco (2023: 205, 207, 219, 220, 221); Slackman (2009).

10. For a report on torching cars by young people in Germany as a form of protest against the establishment, gentrification, and globalization, see Bhatti (2012).

11. On this label/title, see Alhoff and Buciak (2013: 221); Amster (2006: 288); Brown (2020: 238); Holden (2011: C5); Joosse (2012: 75, 88, 89; 2014: 43); Rosenfeld (2006); see also Federman (2013); Fox News (2008/2015); Loadenthal (2017: 3, 5–7, 10, 21); Potter (2010: 24); Resistance Magazine (2009a: 9, 2010: 12); Salter (2011: 212, 214, 218, 221, 227, 228, 229); Selmini and Di Ronco (2023: 220, 221); Smallwood (2005); Vanderheiden (2005: 425, 426). For a reference to the E.L.F. as "among the [F.B.I's] 'highest domestic terrorism priorities'," see Joosse (2021:739 (quoting Robert S. Mueller III, then-director of the F.B.I. (in Janofsky 2006))). On the designation of the Animal Liberation Front (the "A.L.F." or "ALF")—an anonymous, non-hierarchical, cell-structured group that predates and resembles the E.L.F. (discussed in Chapter 2)—as "a top domestic terrorism threat in the U.S.," see Johnston and Johnston (2020: 575); on the F.B.I.'s and Canadian Security Intelligence Service's ("C.S.I.S.") reference to the "Radical Animal Environmental Movement" as a "high-level terrorist threat," see Johnston and Johnston (2017: 735 (citing Berthiaume 2011; Best 2007)).
12. To be clear, in "'Green Harms' as Art Crime, Art Criticism as Environmental Dissent" (Brisman 2011a), I did *not* suggest that animal rights advocates and eco-/environmental activists should *stop* making arguments against animal abuse and cruelty on ethical, moral, or ecophilosophical grounds. I maintained then—and still do—that it is wrong to hurt, maim, or kill animals for art, and that we should hold artists responsible for such actions, as well as gallery and museum curators and directors who exhibit such "works." Rather, I was offering a proposal for activists to expand their arsenal, bolster their artillery, and broaden their critiques with heightened attention to art and aesthetics: "sometimes art that is ethically or morally problematic is also bad art *as art*" (Brisman 2011a: 479 (emphasis in original)). (See Smith (2018: 20) on the benefit of "a plurality of strategies and tactics"; see Brown (2021: 356, 357, 362, 365, 371), Burkett (2016: 17, 36n.240, 42, 43n.285), McIntyre (2021: 225) on "diversity of tactics" and/or "tactical diversity" as a source of strength; and see Malm (2021: 116, 115) on the importance of "'*a diversity and plurality of tactics*'" for social movements—and on the need to "give it appendages" (emphasis in original).)

In "Vandalizing Meaning, Stealing Memory: Artistic, Cultural, and Theoretical Implications of Crime in Galleries and Museums" (Brisman 2011b), I contemplated the theft and vandalism of works of art and argued that how we regard such events should be determined not by their criminality, but by the perpetrators' intent and the effect of the acts on the meaning and memory of the works—especially where the theft or defacement of works of art is intended as an artistic statement in its own right. (In such instances, I maintained, "if a theft or assault is to occur in the name of art, it should be well-conceived, well-executed, conceptually coherent, and aesthetically tight—like any work of art" (Brisman 2011b: 26).) Although that article (Brisman 2011b) did not involve looking to art criticism, art history and aesthetics to explore an *environmental* issue, crime, or harm (as in Brisman 2005a, b, 2011a), it represents a similar kind of inquiry to those pieces: it pondered a crime (or *types* of crime—art theft, vandalism) from the perspective of art criticism and art history, rather than criminology and law. In the last few years (in 2022, in particular), activists have received considerable media coverage of their gluing themselves to masterpieces and/or throwing food at famous works of art in galleries and museums in efforts to call attention to the climate crisis (see, e.g., Adams 2022; AFP 2023; Andreoni 2022; Associated Press 2022c, d, e; Associated Press in Berlin 2022; BBC News 2022b, 2023; Benzine 2022; Boudreau 2022; Braithwaite and Pettersson 2023; Braun 2022; Britton and Associated Press 2022; Burke 2022; Cascone 2022a, b; Christodoulou 2022; Crow 2022; Dafoe 2022a, b; Diaz 2022; Easton 2022; Edwards and Mendonca 2022; Escalante-De Mattei 2022; Feldman 2022; Gayle 2022a, b, c, d; Goldstein 2022; Greenberger 2022; Grenier 2022; Grieshaber 2022; Harris 2022a, b; Hickley 2023; Ho 2022; Holland 2022; Horn-Muller 2022; Jankowicz 2022; Jones 2022; Kateman 2022; Lawson-Tancred 2022a, b, c, d, g; Lim 2022; Lukpat 2022; Mandler 2022; Manson 2024; Marshall 2022a, b; Mauney 2022; McShane 2022a, b; Medina 2022; Monbiot 2022; News Desk 2022a, b, c; Ore 2022; Palumbo 2022; Pawson 2022; Rosa-Aquino 2022; Sackur 2022; Schneider 2022; Scribner 2022; Silverman 2023; Small 2023; Tan 2022; Victor and Cramer 2022; Watson and Stambaugh 2022; Youngs 2022; see generally Lawson-Tancred 2022e, f; News Desk 2023; cf. Lawson-Tancred 2023).

Were I to extend the type of inquiry in Brisman (2011b) to these more recent protests, I would encourage a consideration of, *inter alia*, the artwork selected (e.g., why this particular piece?) and the choice of substances hurled at or smeared on the piece (e.g., why a particular kind of food?). It appears that to some extent, some activists have engaged in this kind of reflection and decision-making (see, e.g., Braun 2022; Escalante-De Mattei 2022; Marshall 2022a; Mauney 2022; Medina 2022; but see Manson 2024).

In a related vein, in November 2023, protesters with the group, Extinction Rebellion NYC, engaged in a creative interruption of Richard Wagner's "Tannhäuser" at the Metropolitan Opera House (part of the Lincoln Center for the Performing Arts) in New York City, deliberately timing their demonstration to "coincide with the main character's declaration that 'love is a spring to be drunk from'" (Mayorquin et al. 2023). As Extinction Rebellion explained in a statement, "contrary to those words spoken on stage, springs are not pure now, because we are in a climate crisis, and our water is contaminated" (quoted in Mayorquin et al. 2023). For a reflection on the protest and the reaction that it elicited, see Barone (2023); for climate activists' interruption of other opera performances, see, e.g., Salazar (2022, 2023).

Such disturbances have not been limited to the arts (and are often not undertaken with the same degree of forethought regarding symbolism and timing). Das (2022) reports that "[d]isrupting high-profile sporting events has become an increasingly common tactic by environmental activists in Europe over the past year, and some have succeeded in drawing attention to their causes in a variety of ways, including attaching themselves to goal posts at Premier League soccer matches; breaching the court at the French Open tennis championships; and blocking the path of cycling's Tour de France." On pitch invaders at English Premier League soccer matches, see, e.g., Andrews (2022), Cole (2022); on breaching courts at tennis championships, see, e.g., Associated Press (2022b), Ng (2023), see also Associated Press (2022a); on obstructing the path of cyclists at the Tour de France, see Reuters (2022). For climate activists' disruption of an auto racing competition, see, e.g., (2022a), Morse (2022), see also Associated Press (2022a); for environmental protests at the U.S. Open Tennis

Championships, see Jiménez and Streeter (2023); for animal rights protesters disruption of National Basketball Association games, see, e.g., Pereles (2022); Polacek (2022); Yohannes (2022).

13. For examples and articulations of green cultural criminology, see, e.g., Brisman (2014, 2015, 2017a, b, c, 2018, 2019a, b, 2020); Brisman et al., (2014); Brisman and South (2012, 2013, 2014, 2015, 2017a, b, c, 2018, 2020, 2022, 2025); Di Ronco (2023); Di Ronco and Allen-Robertson (2021); Di Ronco et al. (2019); Gacek and McClanahan (2021); Kennedy (2021); Komatsubara (2021); Lam et al. (2025); Mazurek (2017); McClanahan (2014, 2019); McClanahan et al. (2017a, b, c); Natali and McClanahan (2017); Natali et al. (2022); Redmon (2018); Schally (2018); Stephens-Griffin (2023).
14. See, e.g., Di Ronco (2023); Di Ronco and Allen-Robertson (2021); Di Ronco and Chiaramonte (2022); Di Ronco et al. (2019); Lee (2021); Pali (2022); Selmini and Di Ronco (2023); Vegh Weis (2021); Watts (2020); see generally Nurik (2022).

    In their article, Selmini and Di Ronco (2023: 197–198) state that while the "[c]riminalization of political dissent is a subject that covers a wide range of interrelated topics and is common to a variety of disciplines, from law to criminology and from sociology to political science[,] [and that] [w]hile theories, processes, and practices of criminalization have been widely analyzed by legal and sociolegal scholars, the criminalization of political dissent in particular has been narrowly studied by social movement scholars and other social scientists—mostly through the lenses of policing." As they submit, "the criminalization of political dissent has been at the margin of the broader, mostly sociolegal, literature on criminalization (Watts 2020) and has been considered almost synonymous with protest policing and therefore predominantly analyzed in studies of police and police violence, rather than more broadly in studies of criminalization processes" (Selmini and Di Ronco 2023: 198). That "criminology has paid virtually no attention to the criminalization of dissent and disobedience (Pali 2022)" (Selmini and Di Ronco 2023: 198) should not be interpreted to mean that the *pattern(s)* and *practice(s)* of criminalizing dissent (by state actors) is or has been sparse or infrequent. As the American activist, Sara Olson, wrote from prison two decades ago, criminalizing dissent through, what were then, "new policies"—particularly, "the Patriot

Act and its Orwellian concept of 'domestic terrorism'"—was "really just persecution as usual in the US" (2006: 268). For Olson (2006: 268–269), "[t]his tactic has been used throughout US history against workers, anti-war, anti-racist, and pro-woman struggles, and it is now being applied against those fighting for animal and earth liberation."

15. Carrabine (2017: 411) explains that "[t]he attention given to images in criminology is a recent development," and that while "there is a rich tradition of research on crime and the media in criminology, specific attention to the role and place of the *image* in crime, in crime control and in criminal justice ha[d] long been lacking" (emphasis in original). This omission began to be addressed in 2014, in a special issue of *Theoretical Criminology*, entitled "Visual Culture and the Iconography of Crime and Punishment," where the late Nicole Rafter offered, in McClanahan's (2021: 5) words, a "foundational definition and description of visual criminology." Noting that criminology "has long preferred the textual to the visual, numerical tables to pictures," Rafter (2014: 128, 129) proclaimed that "[i]f criminology is the study of crime, then visual criminology is the study of ways in which all things visual interact with crime and criminal justice, inventing and shaping one another."

    Given this book's focus on Conceptual art, which employs and relies on *language*, rather than (conventional) visual imagery, and which deemphasizes and often *eschews* the art object, it might seem odd to claim inspiration from *visual* criminology. But I have been motivated by the spirit of visual criminology, which "echoes the activism and opposition to mainstream forms of consensus and classical criminology [located in] radical criminology's [beginnings] in both the UK and the US" (Brown and Carrabine 2019: 192), and have found influential the visual approaches to environmental crime and harm in the work of Natali (2010, 2016), Natali and Budó (2019), Natali and McClanahan (2017, 2020), McClanahan (2021), McClanahan and Linnemann (2018), and Carrabine (2018a), as well as criminological engagement with art and art history (see, e.g., Beirne 2013, 2015, 2018; Beirne and Janssen 2014; Carrabine 2010, 2011a, b, 2012, 2016, 2018a, b, c, 2019, 2021; see generally Carrabine 2014, 2018d).

For an overview of visual criminology, see Brown and Carrabine (2017a) and Carrabine (2017); for a more in-depth exploration of the origins of visual criminology, as well as some of its methodological trends/tendences and contemporary themes, see McClanahan (2021); for an articulation of visual criminology as a critical criminology, see Brown and Carrabine (2019), see also McClanahan (2021), see generally Carrabine 2012); for a primer on visual criminology, offering contributions on formative positions in the development of visual criminology (with attention to its history, theory, and methods), as well as explorations of visual representations of crime and criminal justice institutions, and meditations on ethical issues posed by images of crime and crime control, see Brown and Carrabine (2017b).

16. Despite the fact that Conceptual artists during this period gave precedence to concepts or ideas over more traditional notions of and concerns with aesthetics, materiality, and skillful execution/technique (see Dutton 2009), I very much wanted to include in this book images of the documentation (photographs, written statements) used by the artists I discuss. I thought that doing so would better support the comparison I am seeking to make. Unfortunately, I was unable to obtain permission to publish the desired images other than Richard Long's *A Line Made by Walking* (1967) [Fig. 1].

    Rafter (2014: 128) remarks that when Eamonn Carrabine has encountered difficulties securing permission to publish an image, he has "solved [the problem] by printing a figure with its caption on a blank page, in the center of which runs the announcement: 'Please note that this image could not be reproduced due to restrictions from the rights holder'" (quoting Carrabine 2012: 479). While it was certainly tempting to follow Carrabine's lead in my book—and, perhaps, consistent with some of the spirit of Conceptual art—ultimately, I decided not to adopt his practice. I believe the comparisons and arguments promulgated in this book can stand on their own without reproduced images of documentation(s) of art. That being said, I would encourage readers interested in learning more about the artists I contemplate to conduct simple internet searches of their names and works, which will return results containing links to the images I could not include here.

References

A Quest for Fame Is Said to Lead to Arsons. 2005. *The New York Times* (National edition), August 18: A22. https://www.nytimes.com/2005/08/18/us/a-quest-for-fame-is-said-to-lead-to-arsons.html.

Adams, Abigail. 2022. Why Climate Activists Are Gluing Their Hands to Treasured Works of Art. *People*, October 12. https://people.com/human-interest/why-climate-activists-gluing-their-hands-to-treasured-art/.

AFP. 2023. Suède: des militants écologistes badigeonnenent de peinture un tableau de Monet. *France 24*, June 14. https://www.france24.com/fr/info-en-continu/20230614-suède-des-militantes-écologistes-badigeonnent-de-peinture-un-tableau-de-monet.

Allhoff, Steffen W., and Sebastian K. Buciak. 2013. Action, Reaction, Escalation: Radical Environmental Organisations in the Context of Eco-terrorism and Eco-radicalism. In *Global Environmental Change: New Drivers for Resistance, Crime and Terrorism?*, Achim Maas, Balázs Bodó, Clementine Burnley, Irina Comardicea, and Roger Roffey, ed. pp. 221–242. Baden-Baden, DE: Nomos Verlagsgesellschaft.

Amster, Randall. 2006. Perspectives on Ecoterrorism: Catalysts, Conflations, and Casualties. *Contemporary Justice Review*, 9(3): 287–301. https://doi.org/10.1080/10282580600827991.

Andreoni, Manuela. 2022. Does Throwing Soup Work? Thoughts on Climate Protests. *The New York Times*, December 2. https://www.nytimes.com/2022/12/02/climate/climate-protests-museum-artwork.html.

Andrews, Connor. 2022. Protestor Who Cable Tied Himself to Everton's Goal Explains Why he Stopped Premier League Clash vs Newcastle—in Climate Change Stunt, Warning of 'Mass Starvation' Risk. *Talksport*, March 18. https://talksport.com/football/1066620/protestor-everton-newcastle-climate-change-just-stop-oil-goalpost-tied/.

Archibold, Randal C. 2009. Fast-Moving California Wildfire Threatens Hundreds of Homes. *The New York Times*, September 1: A10.

Armstrong, Sarah, Katherine Biber, and Travis Linnemann. 2020. Editors' Introduction. *Crime Media Culture*, 16(1): 3–5. https://doi.org/10.1177/1741659020906284.

Associated Press. 2004a. National Briefing: Mid-Atlantic: Maryland: Police Have Suspects in Arsons. *The New York Times* (National edition), December 16: A40. https://www.nytimes.com/2004/12/16/us/national-briefing-midatlantic-maryland-police-have-suspects-in-arsons.html.

Associated Press. 2004b. Investigators Consider Race as Possible Motive in Maryland Arson. *The New York Times* (National edition), December 20: A14. https://www.nytimes.com/2004/12/20/us/investigators-consider-race-as-possible-motive-in-maryland-arson.html.

Associated Press. 2005a. National Briefing|Mid-Atlantic: Maryland: Not-Guilty Pleas In Arson Case. *The New York Times* (National edition), January 13: A32. https://www.nytimes.com/2005/01/13/us/national-briefing-mid atlantic-maryland-notguilty-pleas-in-arson-case.html.
Associated Press. 2005b. National Briefing|Mid-Atlantic: Maryland: A Second Suspicion of Arson. *The New York Times* (National edition), September 24: A9. https://www.nytimes.com/2005/09/24/us/national-briefing-mid atlantic-maryland-a-second-suspicion-of-arson.html.
Associated Press. 2014. Protestor in St. Louis Area Charged with Arson. *The New York Times*, December 27. http://www.nytimes.com/aponline/2014/12/27/us/ap-us-killings-by-police-berkeley-protester.html (last visited December 30, 2014). Also available at: https://www.yahoo.com/music/s/protester-st-louis-area-charged-arson-204951252.html (last visited December 30 2014); http://abcnews.go.com/US/wireStory/protester-st-louis-area-charged-arson-27853645 (last visited December 30, 2014).
Associated Press. 2022a. Rafael Nadal Reaches French Open Final to Face Casper Ruud After Alexander Zverev Retires With Ankle Injury. *ESPN*, June 3. https://www.espn.com/tennis/story/_/id/34031734/rafael-nadal-reaches-french-open-final-face-casper-ruud-alexander-zverev-retires-ankle-injury.
Associated Press. 2022b. Environmental activist interrupts the French Open and attaches herself to the net to protest climate change. *Chicago Tribune*, June 4 (originally published June 3, 2022). https://www.chicagotribune.com/2022/06/03/environmental-activist-interrupts-the-french-open-and-attaches-herself-to-the-net-to-protest-climate-change/.
Associated Press. 2022c. UK Climate Protesters Glue Themselves to Gallery Paintings. *Yahoo! News*, July 5. https://www.yahoo.com/news/uk-climate-protesters-glue-themselves-112713659.html?fr=yhssrp_catchall.
Associated Press. 2022d. Climate Protestors Throw Soup on Van Gogh's 'Sunflowers' Painting in London. *National Public Radio* (*N.P.R.*), October 14. https://www.npr.org/2022/10/14/1129098184/van-gogh-sunflowers-soup-climate-protest-london-gallery. Also published as "UK: Climate protesters throw soup on Van Gogh's 'Sunflowers'" on October 14, 2022, at https://apnews.com/article/london-painting-climate-and-environment-b15e0092560b290c04920620b2d7c061.
Associated Press. 2022e. 'Girl with a Pearl Earring' Targeted With Glue and Liquid by Climate Activists. *NBC News*, October 28. https://www.nbcnews.com/news/world/girl-pearl-earring-targeted-glue-liquid-climate-activists-rcna54494.
Associated Press in Berlin. 2022. Climate Activists Throw Black Liquid at Gustav Klimt Painting in Vienna. *The Guardian*, November 15. https://www.theguardian.com/environment/2022/nov/15/climate-activists-throw-black-liquid-at-gustav-klimt-painting-in-vienna.

Austen, Ian, and Vjosa Isai. 2024. Wildfire Roars Through a Canadian Town Popular with Tourists. *The New York Times*, July 25. https://www.nytimes.com/2024/07/25/world/canada/jasper-wildfire-canada-national-park.html?smid=nytcore-ios-share&referringSource=articleShare&sgrp=c-cb.

Badolato, Edward V. 1991. Environmental Terrorism: A Case Study. *Terrorism* [now known as *Studies in Conflict and Terrorism*], 14(4): 237–239. https://doi.org/10.1080/10576109108435882.

Barak, Gregg. 1988. Newsmaking Criminology: Reflections on the Media, Intellectuals, and Crime. *Justice Quarterly*, 5(4): 265–287. https://doi.org/10.1080/07418828800089891.

Barak, Gregg. 1994. *Media, Process, and the Social Construction of Crime: Studies in Newsmaking Criminology*. New York: Garland.

Barak, Gregg. 1996. Media, Discourse, and the O.J. Simpson Trial: An Ethnographic Portrait. In *Representing O.J.: Murder, Criminal Justice, and Mass Culture*, Gregg Barak, ed. pp. 104–122. Guilderland, NY: Harrow and Heston.

Barak, Gregg. 2007. Doing Newsmaking Criminology from Within the Academy. *Theoretical Criminology*, 11(2): 191–207. https://doi.org/10.1177/1362480607075847.

Barak, Gregg. 2017. Newsmaking Criminology. In *The Routledge Companion to Criminological Theory and Concepts*, Avi Brisman, Eamonn Carrabine and Nigel South, ed. pp. 193–196. Abingdon, Oxon, UK, and New York: Routledge.

Barone, Joshua. 2023. Climate Protests Upstage a Met Debut. *The New York Times* (New York edition), December 2: C1. Published online as "Review: Climate Protests Update a Debut at the Met Opera" on December 1, 2023, at https://www.nytimes.com/2023/12/01/arts/music/review-tannhauser-met-opera.html.

Barringer, Felicity. 2004a. Disputed Maryland Subdivision Set Afire. *The New York Times* (National edition), December 7: A18. https://www.nytimes.com/2004/12/07/us/disputed-maryland-subdivision-set-afire.html.

Barringer, Felicity. 2004b. Neighbors of Burned Homes Pained by Suburban Sprawl. *The New York Times* (National edition), December 12: Section 1: 37. https://www.nytimes.com/2004/12/12/us/neighbors-of-burned-homes-pained-by-suburban-sprawl.html.

BBC News. 2019. Australian Bushfires Reach Sydney's Suburbs. *BBC News*, November 12. https://www.bbc.com/news/world-australia-50375407.

BBC News. 2022a. British Grand Prix: Six Charged After F1 Silverstone Track Invasion. *BBC News*, July 5. https://www.bbc.com/news/uk-england-northamptonshire-62050697.

BBC News. 2022b. Climate Protesters Glue Themselves to John Constable Masterpiece. *BBC News*, July 5. https://www.bbc.com/news/uk-england-london-62038615.

BBC News. 2023. Just Stop Oil Activists Fined for Damaging Last Supper. *BBC News*, February 8. https://www.bbc.com/news/uk-england-london-64566239.

Beirne, Piers. 2013. Hogarth's Animals. *Journal of Animal Ethics*, 3(2) [Fall]: 133–162. https://doi.org/10.5406/janimalethics.3.2.0133.

Beirne, Piers. 2015. *Hogarth's Art of Animal Cruelty: Satire, Suffering and Pictorial Propaganda*. Basingstoke, Hampshire, UK: Palgrave Macmillan.

Beirne, Piers. 2018. Raw, Roast or Half-Baked? Hogarth's Beef in *Calais Gate*. *Theoretical Criminology*, 22(3): 426–444. https://doi.org/10.1177/1362480618787174.

Beirne, Piers, and Janine Janssen. 2014. Hunting Worlds Turned Upside Down: Paulus Potter's *Life of a Hunter*. *Tijdschrift over Cultuur & Criminaliteit*, 4(2): 15–28. https://doi.org/10.5553/TCC/221195072014004002002.

Benzine, Vittoria. 2022. Here Is Every Artwork Attacked by Climate Activists This Year, From the 'Mona Lisa' to 'Girl With a Pearl Earring'. *Artnet*, October 31. https://news.artnet.com/art-world/here-is-every-artwork-attacked-by-climate-activists-this-year-from-the-mona-lisa-to-girl-with-a-pearl-earring-2200804.

Berthiaume, Lee. 2011. Federal Website Defines Animal-Rights Groups as Terrorists. *National Post*, October 11. https://nationalpost.com/news/canada/federal-website-defines-animal-rights-groups-as-terrorists.

Best, Steven. 2007. The Animal Enterprise Terrorism Act: New, Improved, and ACLU Approved. *The International Journal of Inclusive Democracy*, 3(3) [July]. https://www.inclusivedemocracy.org/journal/vol3/vol3_no3_best.htm.

Bhatti, Jabenn. 2012. Torching cars is a common protest in Germany. *USA Today*, January 4: 2A. Available at: https://www.pressreader.com/usa/usa-today-us-edition/20120104/287608191604111.

Boudreau, Catherine. 2022. Claude Monet, Larry Fink, and Ted Cruz: Climate Students Grab Headlines but can they get Results? *Business Insider*, October 28. https://www.businessinsider.com/climate-activists-target-claude-monet-larry-fink-and-ted-cruz-2022-10.

Braithwaite, Sharon, and Henrik Pettersson. 2023. Environment Activists Smear Red Paint On Monet Artwork At Stockholm Museum. *CNN*, June 14. https://www.cnn.com/2023/06/14/style/monet-paint-protest-stockholm-climate/index.html.

Braun, Stuart. 2022. Attack on Monet Artwork: German Museums React. *Deutsche Welle*, October 22. https://www.dw.com/en/attack-on-monet-artwork-german-museums-react/a-62927968.

Brisman, Avi. 2005a. The Aesthetics of Wind Energy Systems. *New York University Environmental Law Journal*, 13(1): 1–133.
Brisman, Avi. 2005b. Letter to the Editor. *Creative Loafing* (Atlanta), August 18–24: 8.
Brisman, Avi. 2008. Crime-Environment Relationships and Environmental Justice. *Seattle Journal for Social Justice*, 6(2): 727–817. https://digitalcommons.law.seattleu.edu/cgi/viewcontent.cgi?article=1138&context=sjsj.
Brisman, Avi. 2011a. "Green Harms" as Art Crime, Art Criticism as Environmental Dissent. *Journal of Contemporary Criminal Justice*, 27(4): 465–499. https://doi.org/10.1177/1043986211418888.
Brisman, Avi. 2011b. Vandalizing Meaning, Stealing Memory: Artistic, Cultural, and Theoretical Implications of Crime in Galleries and Museums. *Critical Criminology: An International Journal*, 19(1): 15–28. https://doi.org/10.1007/s10612-010-9104-x.
Brisman, Avi. 2014. Of Theory and Meaning in Green Criminology. *International Journal for Crime, Justice and Social Democracy*, 3(2): 22–35. https://doi.org/10.5204/ijcjsd.v3i2.173; https://www.crimejusticejournal.com/article/view/723.
Brisman, Avi. 2015. Environment and Conflict: A Typology of Representations. In *Environmental Crime and Social Conflict: Contemporary and Emerging Issues*, Avi Brisman, Nigel South, and Rob White, ed. pp. 285–311. Surrey, UK: Ashgate.
Brisman, Avi. 2017a. An Epilogue to the Book, not an Elegy for the Earth. In *Environmental Crime in Latin America: The Theft of Nature and the Poisoning of the Land,* David Rodríguez Goyes, Hanneke Mol, Avi Brisman, and Nigel South, ed. pp. 297–301. London: Palgrave Macmillan.
Brisman, Avi. 2017b. On Narrative and Green Cultural Criminology. *International Journal for Crime, Justice and Social Democracy*, 6(2): 64–77. https://doi.org/10.5204/ijcjsd.v6i2.347; https://www.crimejusticejournal.com/article/view/868.
Brisman, Avi. 2017c. Representations of Environmental Crime and Harm: A Green-Cultural Criminological Perspective on *Human-Altered Landscapes*. In *The Routledge International Handbook of Visual Criminology*, Michelle Brown and Eamonn Carrabine, ed. pp. 523–539. London and New York: Routledge.
Brisman, Avi. 2018. Representing the "Invisible Crime" of Climate Change in an Age of Post-Truth. *Theoretical Criminology*, 22(3): 468–491. https://doi.org/10.1177/1362480618787168.
Brisman, Avi. 2019a. The Fable of *The Three Little Pigs*, Climate Change and Green Cultural Criminology. *International Journal for Crime, Justice and Social Democracy*, 8(1): 46–69. https://doi.org/10.5204/ijcjsd.v8i1.952; https://www.crimejusticejournal.com/article/view/952.

Brisman, Avi. 2019b. Stories of Environmental Crime, Harm and Protection: Narrative Criminology and Green Cultural Criminology. In *The Emerald Handbook of Research in Narrative Criminology*, Jennifer Fleetwood, Lois Presser, Sveinung Sandberg, and Thomas Ugelvik, ed. pp. 153–172. Bingley, West Yorkshire, UK: Emerald Group Publishing.

Brisman, Avi. 2020. Cultural Criminology and Narrative Criminology's Shared Interests. More Than Just Criminological Verstehen. *Tijdschrift over Cultuur & Criminaliteit,* (10)3: 14–34. https://doi.org/10.5553/TCC/221195072020010003002.

Brisman, Avi, Bill McClanahan, and Nigel South. 2014. Toward a Green-Cultural Criminology of "The Rural." *Critical Criminology: An International Journal*, 22(4): 479–494. https://doi.org/10.1007/s10612-014-9250-7.

Brisman, Avi, and Nigel South. 2012. Proposing a Green-Cultural Criminology: a new perspective for 21st century criminology. In *Ecological Crime and Environmental Protection—Multidisciplinary Perspectives*, Gorazd Meško, Andrej Sotlar and Katja Eman, ed. pp. 17–35. Llubljana, Slovenia: Faculty of Criminal Justice, University of Maribor. Published as: Brisman, Avi, in Nigel South. 2012. Predlog ekološko-kulturne kriminologije: nov pogled za kriminologijo 21. Stoletja. *Ekološka kriminaliteta in varovanje okolja – multidisciplinarne perspektive.* Gorazd Meško, Andrej Sotlar and Katja Eman (uredniki). Pp.17–35. Ljubljana: Fakuleta za varnostne vede, Univerza v Mariboru.

Brisman, Avi, and Nigel South. 2013. A Green-Cultural Criminology: An Exploratory Outline. *Crime Media Culture*, 9(2): 115–135. https://doi.org/10.1177/1741659012467026.

Brisman, Avi, and Nigel South. 2014. *Green Cultural Criminology: Constructions of Environmental Harm, Consumerism, and Resistance to Ecocide.* Abingdon, Oxon, UK, and New York: Routledge.

Brisman, Avi, and Nigel South. 2015. 'Life Stage Dissolution', Infantilization and Anti-social Consumption: Implications for De-responsibilization, Denial and Environmental Harm. *Young—Nordic Journal of Youth Research*, 23(3): 209–221. https://doi.org/10.1177/1103308815584876.

Brisman, Avi, and Nigel South. 2017a. Consumer Technologies, Crime and Environmental Implications. In *The Routledge Handbook of Technology, Crime and Justice*, M.R. McGuire and Thomas J. Holt, ed. pp. 310–324. London and New York: Routledge.

Brisman, Avi, and Nigel South. 2017b. Criminología verde cultural [Green cultural criminology]. In *Introducción a la criminología verde. Conceptos para nuevos horizontes y diálogos socioambientales* [*Introduction to Green Criminology: Concepts for New Horizons and Socio-Environmental Dialogues*], Hanneke Mol, David Rodríguez Goyes, Nigel South, and Avi Brisman, ed.

pp. 97–127. Bogotá, Colombia: Editorial Temis S.A. and Universidad Antonio Nariño, Fondo Editorial.

Brisman, Avi, and Nigel South. 2017c. Green Cultural Criminology, Intergenerational (In)equity and "Life Stage Dissolution." In *Greening Criminology in the 21st Century: Contemporary Debates and Future Directions in the Study of Environmental Harm*, Matthew Hall, Jennifer Maher, Angus Nurse, Gary Potter, Nigel South, and Tanya Wyatt, ed. pp. 219–232. Abingdon, Oxon, UK: Routledge.

Brisman, Avi, and Nigel South. 2018. Green Cultural Criminology. In *Routledge Handbook of Critical Criminology*, 2nd edn, Walter S. DeKeseredy and Molly Dragiewicz, ed. pp. 132–142. London and New York: Routledge.

Brisman, Avi, and Nigel South. 2020. Toward a Green Cultural Criminology of the South. In *Routledge International Handbook of Green Criminology*, 2nd edn, Avi Brisman and Nigel South, ed. pp. 624–637. Abingdon, Oxon, UK, and New York: Routledge.

Brisman, Avi, and Nigel South. 2022. Criminologia verde cultural: Explorando a literature e narrativas culturais indígenas da América Latina [Green cultural criminology: Exploring literary and Indigenous cultural narratives from Latin America]. In *Introdução à criminologia verde: Perspectivas críticas, decoloniais e do Sul* [Introduction to Green Criminology: Southern, decolonial and critical perspectives], Marília de Nardin Budó, David Rodríguez Goyes, Lorenzo Natali, Ragnhild Sollund and Avi Brisman, ed. pp. 103–115. São Paulo, Brasil: Tirant lo Blanch.

Brisman, Avi, and Nigel South. 2025. *Monstrous Nature and Representations of Environmental Harms: A Green Cultural Criminological Perspective.* Philadelphia, PA: Temple University Press.

Britton, Bianca, and The Associated Press. 2022. Man in Wig Throws Cake at Mona Lisa in Climate Protest Stunt. *NBC News*, May 30. https://www.nbcnews.com/news/world/man-wig-throws-cake-mona-lisa-climate-protest-stunt-rcna31094.

Brown, Alleen. 2019. The Green Scare: How a Movement That Never Killed Anyone Became the FBI's No. 1 Domestic Terrorism Threat. *The Intercept,* March 23. Available at: https://theintercept.com/2019/03/23/ecoterrorism-fbi-animal-rights/.

Brown, Joseph M. 2020. Notes to the Underground: Credit Claiming and Organizing in the Earth Liberation Front. *Terrorism and Political Violence*, 32(2): 237–256. https://doi.org/10.1080/09546553.2017.1364637. Published online on September 7, 2017.

Brown, Joseph M. 2021. Civil Disobedience, Sabotage, and Violence in US Environmental Activism. In *The Oxford Handbook of Comparative Environmental Politics*, Jeannie Sowers, Stacy D. VanDeveer, and Erika Weinthal, ed.

pp. 356–374. Online edition, Oxford Academic, https://doi.org/10.1093/oxfordhb/9780197515037.013.34.

Brown, Michelle. 2014. Visual Criminology and Carceral Studies: Counter-Images in the Carceral Age. *Theoretical Criminology*, 18(2): 176–197. https://doi.org/10.1177/1362480613508426.

Brown, Michelle, and Eamonn Carrabine. 2017a. Introducing Visual Criminology. In *Routledge International Handbook of Visual Criminology*, Michelle Brown and Eamonn Carrabine, ed. pp. 1–9. Abingdon, Oxon, UK, and New York: Routledge.

Brown, Michelle, and Eamonn Carrabine, ed. 2017b. *Routledge International Handbook of Visual Criminology*. Abingdon, Oxon, UK, and New York: Routledge.

Brown, Michelle, and Eamonn Carrabine. 2019. The Critical Foundations of Visual Criminology: The State, Crisis, and the Sensory. *Critical Criminology: An International Journal*, 27(1): 191–205. https://doi.org/10.1007/s10612-019-09439-7.

Burke, Olivia. 2022. Mash for Brains: Eco-idiots Hurl Mashed Potatoes over £100 Million Monet Painting & Glue Themselves to Wall in Daft Stunt. *The Sun*, October 23 [updated October 24, 2022]. https://www.thesun.co.uk/news/20199054/eco-idiots-hurl-mash-potatoes-monet-painting-glue-themselves-floor/.

Burkett, Maxine. 2016. Climate Disobedience. *Duke Environmental Law & Policy Forum*, 27(1) [Fall]: 1–50. https://scholarship.law.duke.edu/delpf/vol27/iss1/1/.

Carrabine, Eamonn. 2010. Imagining Prison: Culture, History, Space. *Prison Service Journal*, 187: 15–22.

Carrabine, Eamonn. 2011a. The Iconography of Punishment: Execution Prints and the Death Penalty. *The Howard Journal of Crime and Justice*, 50(5) [December]: 452–464. https://doi.org/10.1111/j.1468-2311.2011.00688.x.

Carrabine, Eamonn. 2011b. Images of Torture: Culture, Politics and Power. *Crime Media Culture*, 7(1) [April]: 245–260. https://doi.org/10.1177/1741659011404418.

Carrabine, Eamonn. 2012. Just Images: Aesthetics, Ethics and Visual Criminology. *The British Journal of Criminology*, 52(3): 463–489. https://doi.org/10.1093/bjc/azr089.

Carrabine, Eamonn. 2014. Seeing Things: Violence, Voyeurism and the Camera. *Theoretical Criminology*, 18(2) [May]: 134–158. https://doi.org/10.1177/1362480613508425.

Carrabine, Eamonn. 2016. Picture This: Criminology, Image and Narrative. *Crime Media Culture*, 12(2) [August]: 253–270. https://doi.org/10.1177/1741659016637640.

Carrabine, Eamonn. 2017. Visual Criminology. In *The Routledge Companion to Criminological Theory and Concepts*, Avi Brisman, Eamonn Carrabine and Nigel South, ed. pp. 411–414. Abingdon, Oxon, UK, and New York: Routledge.

Carrabine, Eamonn. 2018a. Geographies of Landscape: Representation, Power and Meaning. *Theoretical Criminology*, 22(3) [August]: 445–467. https://doi.org/10.1177/1362480618787172.

Carrabine, Eamonn. 2018b. Punishment in the Frame: Rethinking the History and Sociology of Art. *The Sociological Review*, 66(3): 559–576. https://doi.org/10.1177/0038026117705031.

Carrabine, Eamonn. 2018c. Reading a "Titian": Visual Methods and the Limits of Interpretation. *Deviant Behavior*, 39(4): 525–538. https://doi.org/10.1080/01639625.2017.1407113.

Carrabine, Eamonn. 2018d. Traces of Violence: Representing the Atrocities of War. *Criminology & Criminal Justice*, 18(5): 631–646. https://doi.org/10.1177/1748895818789448.

Carrabine, Eamonn. 2019. Reading Pictures: Piranesi and Carceral Landscapes. In *The Emerald Handbook of Narrative Criminology*, Jennifer Fleetwood, Lois Presser, Sveinung Sandberg, and Thomas Ugelvik, ed. pp. 197–216. Bingely, UK: Emerald Publishing Limited.

Carrabine, Eamonn. 2021. Low life: William Hogarth, Visual Culture and Sociologies of Art. *The British Journal of Sociology*, 72(4) [September]: 909–929. https://doi.org/10.1111/1468-4446.12849; https://onlinelibrary.wiley.com/.

Cascone, Sarah. 2022a. Can the Climate Protesters Gluing Themselves to Frames of Masterpieces Damage Irreplaceable Art? We Asked the Experts. *Artnet*, August 28. https://news.artnet.com/art-world-archives/museum-climate-protesters-glue-risk-2165891.

Cascone, Sarah. 2022b. A Climate Activist Just Superglued His Head to Vermeer's 'Girl With a Pearl Earring'. *Artnet*, October 27. https://news.artnet.com/art-world-archives/a-climate-activist-just-superglued-his-head-to-vermeers-girl-with-a-pearl-earring-2199859.

Chester, Lynne. 2020. The 2019–2020 Australian Bushfires: A Potent Mix of Climate Change, Problematisation, Indigenous Disregard, A Fractured Federation, Volunteerism, Social Media, and More. *Review of Evolutionary Political Economy*, 1: 245–264. https://doi.org/10.1007/s43253-020-00019-y.

Chibnall, Steve. 1977. *Law-and-Order News*. London: Tavistock.

Christodoulou, Holly. 2022. Stroke of Stupidity: Activist Yobs Vandalise Van Gogh's Sunflowers and Douse Masterpiece With Soup in National Gallery Rampage. *The Sun*, October 14. https://www.thesun.co.uk/news/20107128/van-gogh-sunflowers-vandalised-protesters/.

Clark, Wesley K., and Kal Raustiala. 2007. Why Terrorists Aren't Soldiers. *The New York Times*, August 8: A19. https://www.nytimes.com/2007/08/08/opinion/08clark.html.

Cohen, Stanley, and Jock Young. 1973. *The Manufacture of News: Deviance, Social Problems and the Media*. Beverly Hills, CA: Sage Publications.

Cole, Jackson. 2022. Man Who Tied Himself to Goalpost Arrested After Everton's Dramatic Victory Over Newcastle was Halted by 'Just Stop Oil' Protest and Eddie Howe Admits: 'I've Never Seen Anything Like It'. *Talksport*, March 18. https://talksport.com/football/1066439/stoppage-everton-newcastle-man-ties-himself-goal-post/.

Combs, Cindy C. 2006. *Terrorism in the Twenty-First Century*, 4th edn. Upper Saddle River, NJ: Pearson Prentice Hall.

Cousins, Sophie. 2020. Bushfires Expose Weaknesses in Australia's Health System. *The Lancet*, 395(10219) [January 18]: 175–176. https://doi.org/10.1016/S0140-6736(20)30096-9; https://www.thelancet.com/journals/lancet/article/PIIS0140-6736(20)30096-9/abstract.

Crow, Kelly. 2022. How Museums Aim to Stop Food-Throwing, Climate-Change Protestors. *The Wall Street Journal*, November 7. https://www.wsj.com/articles/museum-food-throwing-protesters-11667829440.

Curra, John. 2017. *The Relativity of Deviance*, 4th edn. Thousand Oaks, CA, and London: Sage.

Dafoe, Taylor. 2022a. Climate Protestors Glued Themselves to a Van Gogh Painting at the Courtauld, Demanding That Museums Join Their 'Civil Resistance'. *Artnet*, June 30. https://news.artnet.com/art-world-archives/just-stop-oil-van-gogh-2139525.

Dafoe, Taylor. 2022b. Activists in Italy Glued Themselves to an Umberto Boccioni Sculpture in Milan to Draw Attention to Climate Change. *Artnet*, August 2. https://news.artnet.com/art-world-archives/climate-activists-italy-glued-themselves-boccioni-2154799.

Das, Andrew. 2022. A climate activist set himself on fire on court at the Laver Cup. *The New York Times*, September 23. Available at: https://www.nytimes.com/2022/09/23/sports/tennis/laver-cup-fire-protest.html.

Diaz, Jaclyn. 2022. Climate Protesters in England Glued Themselves to a Copy of 'The Last Supper'. *National Public Radio* (*N.P.R.*), July 5. https://www.npr.org/2022/07/05/1109848464/climate-protest-uk-art-last-supper#:~:text=A%20group%20of%20at%20least%20five%20activists%20from,them%20that%20one%20of%20them%20will%20betray%20him.

Di Ronco, Anna. 2023. *Policing Environmental Protest: Power and Resistance in Pandemic Times*. Bristol, UK: Bristol University Press.

Di Ronco, Anna, and James Allen-Robertson. 2021. Representations of Environmental Protest on the Ground and in the Cloud: The NOTAP Protests in Activist Practice and Social Visual Media. *Crime Media Culture*, 17(3): 375–399. https://doi.org/10.1177/1741659020953889.

Di Ronco, Anna, James Allen-Robertson, and Nigel South. 2019. Representing Environmental Harm and Resistance on Twitter: The Case of the TAP Pipeline in Italy. *Crime Media Culture*, 15(1): 143–168. https://doi.org/10.1177/1741659018760106.

Di Ronco, Anna, and Xenia Chiaramonte. 2022. Harm to Knowledge: Criminalising Environmental Movements Speaking Up Against Megaprojects. In *The Palgrave Handbook of Restorative Justice*, B. Pali, M. Fosyth, and F. Tepper, ed. pp. 421–447. Cham, CH: Palgrave Macmillan.

Dutton, Denis. 2009. Has Conceptual Art Jumped the Shark Tank. *The New York Times* (New York edition), October 16: A27. Published online as "Has Conceptual Art Jumped the Shark Tank" on October 15, 2009, at https://www.nytimes.com/2009/10/16/opinion/16dutton.html.

Easton, Kaitlin. 2022. Oil Protestors Glue Themselves to Painting at Scots are Gallery. *Daily Record* (Scotland), June 29. https://www.dailyrecord.co.uk/news/scottish-news/protestors-glue-glasgow-art-gallery-27359644.

Edwards, Christian, and Duarte Mendonca. 2022. Fossil Fuel Protesters Charged After Tomato Soup Thrown on Van Gogh's 'Sunflowers' in London Gallery. *CNN.com*, October 15. https://www.cnn.com/style/article/oil-protest-van-gogh-sunflower-soup-intl-scli-gbr/index.html.

Escalante-De Mattei, Shanti. 2022. Climate Activists Glued Themselves to One of the Vatican's Most Valuable Artworks. *ARTnews*, August 19. https://www.artnews.com/art-news/news/climate-change-protest-glue-to-vatican-sculpture-1234636978/.

Evershed, Nick, and Andy Ball. 2019. How Australia's Bushfires Spread: Mapping the East Coast Fires. *The Guardian*, November 21. https://www.theguardian.com/news/datablog/2019/nov/21/how-australias-bushfires-spread-mapping-the-nsw-and-queensland-fires.

Farrer, Ben, and Graig R. Klein. 2022. How Radical Environmental Sabotage Impacts US Elections. *Terrorism and Political Violence*, 34(2): 218–239. https://doi.org/10.1080/09546553.2019.1678468.

Federman, Adam. 2013. Climate Activists, Eco-Terrorism, and the Green Scare. *Utne*. September/October. http://www.utne.com/politics/climate-activists-zm0z13sozlin.aspx#axzz3N9uska3M (last visited August 15, 2024; on file with author).

Feldman, Ella. 2022. Climate Activists Throw Soup on Vincent van Gogh's 'Sunflowers' to Protest Fossil Fuels. *Smithsonian*, October 17. https://www.smithsonianmag.com/smart-news/climate-activists-throw-soup-on-vincent-van-goghs-sunflowers-to-protest-fossil-fuels-180980958/.

Ferrell, Jeff. 1998. Criminalising Popular Culture. In *Popular Culture, Crime and Justice*, Frankie Y. Bailey and Donna C. Hale, ed. pp. 71–86. Belmont, CA: West/Wadsworth.

Finlay, Christopher J. 2009. How to do Things with the Word 'Terrorist.' *Review of International Studies*, 35(4) [October]: 751–774. http://www.jstor.org/stable/40588072.

Fox News. 2008. FBI: Eco-Terrorism Remains No. 1 Domestic Terror Threat. *Fox News*, March 31 [updated January 14, 2015]. https://www.foxnews.com/story/fbi-eco-terrorism-remains-no-1-domestic-terror-threat.

Furedi, Frank. 2009. Terrorism and the Politics of Fear. In *Criminology*, 2nd edn, Chris Hale, Keith Hayward, Azrini Wahidin, and Emma Wincup, ed. pp. 343–361. Oxford: Oxford University Press.

Gacek, James, and Bill McClanahan. 2021. Learning about the Illegal and the "Lawful yet Awful": Progressing Public Criminology in the University Classroom. *Journal of Criminal Justice Education*, 32(4): 495–512. https://doi.org/10.1080/10511253.2021.1916051.

Gans, Herbert J. 1979. *Deciding What's News: A Study of CBS Evening News, NBC Nightly News, Newsweek, and Time*. New York: Pantheon Books.

Gately, Gary. 2004a. Arson Report Says Suspect Felt No Caring From Bosses. *The New York Times* (National edition), December 18: A13. https://www.nytimes.com/2004/12/18/us/arson-report-says-suspect-felt-no-caring-from-bosses.html.

Gately, Gary. 2004b. 2 More Men Are Arrested at Maryland Development. *The New York Times* (National edition), December 21: A22. https://www.nytimes.com/2004/12/21/us/2-more-men-are-arrested-in-arson-at-maryland-development.html.

Gately, Gary. 2004c. As Investigation Widens, Arson Is Linked to Maryland Gang. *The New York Times* (National edition), December 22: A20. https://www.nytimes.com/2004/12/22/us/as-investigation-widens-arson-is-linked-to-maryland-gang.html.

Gately, Gary. 2004d. National Briefing|Mid-Atlantic: Maryland: No Bail in Arson Case. *The New York Times* (National edition), December 24: A15. https://www.nytimes.com/2004/12/24/us/national-briefing-midatlantic-maryland-no-bail-in-arson-case.html.

Gately, Gary. 2005a. Federal Jury Indicts 5 in Subdivision Fires in Maryland. *The New York Times* (National edition), January 4: A13. https://www.nytimes.com/2005/01/04/us/federal-jury-indicts-5-in-subdivision-fires-in-maryland.html.

Gately, Gary. 2005b. Man Pleads Guilty to Arson Spree in Maryland. *The New York Times* (National edition), April 29: A17. https://www.nytimes.com/2005/04/29/us/man-pleads-guilty-to-arson-spree-in-maryland.html.

Gately, Gary. 2005c. Man Convicted in Costly Maryland Arsons. *The New York Times* (National edition), September 3: A11. https://www.nytimes.com/2005/09/03/us/man-convicted-in-costly-maryland-arsons.html.

Gately, Gary. 2005d. Pall of Racism Remains Over Neighborhood Repaired After Arson. *The New York Times* (National edition), October 6: A16. https://www.nytimes.com/2005/10/06/us/pall-of-racism-remains-over-neighborhood-repaired-after-arson.html.

Gately, Gary. 2005e. Ringleader in Arsons Gets Maximum Sentence. *The New York Times* (National edition), December 6: A18. https://www.nytimes.com/2005/12/06/us/ringleader-in-arsons-gets-the-maximum-sentence.html.

Gately, Gary. 2006a. National Briefing | Mid-Atlantic: Maryland: Jury Deadlocks in Arson Trial. *The New York Times* (National edition), March 17: A21. https://www.nytimes.com/2006/03/17/us/national-briefing-midatlantic-maryland-jury-deadlocks-in-arson-trial.html.

Gately, Gary. 2006b. National Briefing| Mid-Atlantic: Maryland: Arson Guilty Plea. *The New York Times*, June 6: A19.

Gayle, Damien. 2022a. Just Stop Oil activists glue themselves to Turner painting frame in Manchester. *The Guardian*, July 1. https://www.theguardian.com/environment/2022/jul/01/just-stop-oil-activists-glue-themselves-to-turner-painting-in-manchester.

Gayle, Damien. 2022b. Just Stop Oil campaigners glue themselves to Da Vinci copy in Royal Academy. *The Guardian*, July 5. https://www.theguardian.com/artanddesign/2022/jul/05/just-stop-oil-campaigners-glue-themselves-to-da-vinci-copy-in-royal-academy.

Gayle, Damien. 2022c. Climate activists in Italy glue themselves to Botticelli painting. *The Guardian*, July 22. https://www.theguardian.com/environment/2022/jul/22/climate-activists-in-italy-glue-themselves-to-botticelli-painting.

Gayle, Damien. 2022d. Just Stop Oil activists throw soup at Van Gogh's Sunflowers. *The Guardian*, October 14. https://www.theguardian.com/environment/2022/oct/14/just-stop-oil-activists-throw-soup-at-van-goghs-sunflowers.

Gergis, Joëlle. 2024. Wildfires. In *The Climate Book*, compiled/created by Greta Thunberg, 96–98. New York: Penguin Books.

Gleick, Peter H. 2006. Water and Terrorism. *Water Policy*, 8(6): 481–503. https://doi.org/10.2166/wp.2006.035.

Glick, Daniel. 2003. *Powder Burn: Arson, Money, and Mystery on Vail Mountain.* New York: PublicAffairs.

Goldstein, Caroline. 2022. 'We Have Been Deeply Shaken': 92 Museum Leaders Release Statement Addressing Climate Actions That Target Works of Art.

Artnet, November 9. https://news.artnet.com/art-world/icom-statement-museum-climate-action-vandalism-2207520.

Graber, Doris. 1980. *Crime News and the Public*. New York: Praeger.

Greek, Cecil, and Deborah B. Henry. 1997. Criminal Justice Resources on the Internet. *Journal of Criminal Justice Education*, 8(1): 91–99. https://doi.org/10.1080/10511259700083981.

Greenberger, Alex. 2022. Climate Activists Throw Mashed Potatoes at $110 M. Monet Painting in Germany. *ARTnews*, October 23. https://www.artnews.com/art-news/news/monet-meules-museum-barberini-climate-activists-protest-1234644090/.

Grenier, Elizabeth. 2022. Climate activists target famous artworks. *Deutsche Welle*, July 5. https://www.dw.com/en/uk-climate-activists-glue-themselves-to-masterpieces/a-62366285.

Grieshaber, Kirsten. 2022. German Art Group Criticizes Climate Protesters for Using Art. *Associated Press*, August 25. https://apnews.com/article/germany-painting-climate-and-environment-4173545ff32525e4ee8e77cc4fcb4042.

Hagmann, R.K., P. F. Hessburg, S. J. Prichard, N. A. Povak, P. M. Brown, P. Z. Fulé, R. E. Keane, E. E. Knapp, J. M. Lydersen, K. L. Metlen, M. J. Reilly, A. J. Sánchez Meador, S. L. Stephens, J. T. Stevens, A. H. Taylor, L. L. Yocom, M. A. Battaglia, D. J. Churchill, L. D. Daniels, D. A. Falk, P. Henson, J. D. Johnston, M. A. Krawchuk, C. R. Levine, G. W. Meigs, A. G. Merschel, M. P. North, H. D. Safford, T. W. Swetnam, A. E. M. Waltz. 2021. Evidence for Widespread Changes in the Structure, Composition, and Fire Regimes of Western North American Forests. *Ecological Applications*, 31(8) [December]: e02431. https://doi.org/10.1002/eap.2431.

Hankin, Lacey E., Chad T. Anderson, Garrett J. Dickman, Parker Bevington, and Scott L. Stephens. 2023. How Forest Management Changed the Course of the Washburn Fire and the Fate of Yosemite's Giant Sequoias (*Sequoiadendron giganteum*). *Fire Ecology*, 19(40): [1–13]. https://doi.org/10.1186/s42408-023-00202-6; https://doi.org/10.1186/s42408-023-00202-6#citeas.

Harris, Gareth. 2022a. Activists Glue Themselves to Painting in Glasgow Museum to Protest Art World's Fossil Fuel Use. *The Art Newspaper*, June 30. https://www.theartnewspaper.com/2022/06/30/environmental-activists-glue-themselves-to-a-painting-in-glasgow-museum-in-protest-against-art-world-complicity-in-fossil-fuel-industry.

Harris, Gareth. 2022b. In Germany, Climate Activists Throwed Mashed Potato at Most Expensive Money Painting Sold at Auction. *The Art Newspaper*, October 24. https://www.theartnewspaper.com/2022/10/24/in-germany-climate-activists-throw-mashed-potato-on-the-most-expensive-monet-painting-sold-at-auction.

Harvey, Fiona. 2023. Scientists Deliver 'Final Warning' on Climate Crisis: Act Now or it's Too Late. *The Guardian*, March 20. https://www.theguardian.com/environment/2023/mar/20/ipcc-climate-crisis-report-delivers-final-warning-on-15c.

Hays, Scott, Michael Esler, and Carol Hays. 1996. Radical Environmentalism and Crime. In *Environmental Crime and Criminality: Theoretical and Practical Issues*, Sally M. Edwards, Terry D. Edwards, and Charles B. Fields, ed. pp. 163–181. New York and London: Garland Publishing.

Hayward, Keith, and Jonathan Ilan. 2017. In *The Routledge Companion to Criminological Theory and Concepts*, Avi Brisman, Eamonn Carrabine and Nigel South, ed. pp. 262–266. Abingdon, Oxon, UK, and New York: Routledge.

Henry, Stuart. 1994. Newsmaking Criminology as Replacement Discourse. In *Media, Process, and the Social Construction of Crime: Studies in Newsmaking Criminology*, Gregg Barak, ed. pp. 287–318. New York: Garland.

Hessburg, Paul F., Susan J. Prichard, R. Keala Hagmann, Nicholas. A. Povak, and Frank K. Lake. 2021. Wildfire and Climate Change Adaptation of Western North American Forests: A Case for Intentional Management. *Ecological Applications,* 31(8) [December]: e02432. https://doi.org/10.1002/eap.2432.

Hickley, Catherine. 2023. Monet Painting Attacked with Red Paint and Glue in Stockholm is Undamaged, Nationalmusem Says. *The Art Newspaper*, June 15. https://www.theartnewspaper.com/2023/06/15/monet-attacked-with-red-paint-and-glue-in-stockholm-is-undamaged-nationalmuseum-says.

Ho, Karen K. 2023. Climate Activists Smear Red Paint on Monet Painting at National Museum in Sweden. *ARTnews*, June 14. https://www.artnews.com/art-news/news/climate-activists-monet-protest-sweden-national-museum-1234671759/.

Holden, Stephen. 2011. Crimes Against Property, as Protests. *The New York Times* (New York edition), June 22: C5. Published online as "Crimes Against Property, as Protests" on June 21, 2011, at https://www.nytimes.com/2011/06/22/movies/if-a-tree-falls-documentary-on-earth-liberation-front.html?searchResultPosition=1.

Holland, Oscar. 2022. Climate protesters glue themselves to 200-year-old masterpiece. *CNN.com*, July 5. https://www.cnn.com/style/article/just-stop-oil-hay-wain-london-climate/index.html.

Hope, Matt. 2020. Australia burning. *The Lancet Planetary Health*, 4(1) [January]: e-12-e13. https://doi.org/10.1016/S2542-5196(20)30006-1; https://www.thelancet.com/pdfs/journals/lanplh/PIIS2542-5196(20)30006-1.pdf.

Horn-Muller, Ayurella. 2022. Climate Activists Divided on Souping Art. *Axios*, October 26. https://www.axios.com/2022/10/26/climate-activists-protests-soup-paintings.

Illing, Sean. 2021. The Case for a More Radical Climate Movement: Author Andreas Malm on the Failures of Climate Activism and the Need for Escalation. *Vox*, October 1. https://www.vox.com/vox-conversations-podcast/22691428/vox-conversations-climate-change-andreas-malm.

Jämte, Jan, and Rune Ellefsen. 2020. The Consequences of Soft Repression. *Mobilization: An International Journal*, 25(3): 383–404. https://doi.org/10.17813/1086-671X-25-3-383.

Jankowicz, Mia. 2022. Climate Activists Dumped Tomato Soup Over a Famed Van Gogh Painting in an Anti-Oil Protest. *Business Insider*, October 14. https://www.businessinsider.com/activists-throw-tomato-soup-van-gogh-painting-climate-protest-2022-10.

Janofsky, Michael. 2006. 11 Indicted in 17 Cases of Sabotage in West. *The New York Times* (National edition), January 21: A9. Published online as "11 Indicted in Cases of Environmental Sabotage" on January 21, 2006, at https://www.nytimes.com/2006/01/21/politics/11-indicted-in-cases-of-environmental-sabotage.html.

Jiménez, Jesus, and Kurt Streeter. 2023. In Second Set, Protesters Cause Bewildering Delay. *The New York Times* (New York edition), September 8: B9. Published online as "Protest Stops U.S. Open Semifinal Match" on September 7, 2023, at https://www.nytimes.com/2023/09/07/sports/tennis/protest-stops-us-open-semifinal-match.html.

Johnston, Genevieve, and Matthew S. Johnston. 2017. 'We Fight For All Living Things': Countering Misconceptions About the Radical Animal Liberation Movement. *Social Movement Studies*, 16(6): 735–751. https://doi.org/10.1080/14742837.2017.1319268.

Jolly, Vik, Kate Selig, and Rachel Parsons. 2024. As California Fires Burn, Some Residents Begin to Mourn Lost Homes. *The New York Times* (New York edition), September 15: A20. Published online as "'Nothing Left': As California Fires Burn, Some Residents Begin to Mourn Lost Homes" on September 14, 2024, at https://www.nytimes.com/2024/09/14/us/california-fires-homes.html?smid=nytcore-ios-share&referringSource=articleShare&sgrp=c-cb&ngrp=mnp&pvid=066C465F-2A9A-498B-9844-657FADA0B441.

Jones, Sam. 2022. Climate Activists Throw Mashed Potatoes at Monet Work in Germany. *The Guardian*, October 23. https://www.theguardian.com/environment/2022/oct/23/climate-activists-mashed-potato-monet-potsdam-germany.

Joosse, Paul. 2007. Leaderless Resistance and Ideological Inclusion: The Case of the Earth Liberation Front. *Terrorism and Political Violence*, 19(3): 351–368. https://doi.org/10.1080/09546550701424042.

Joosse, Paul. 2012. Elves, Environmentalism, and "Eco-terror": Leaderless Resistance and Media Coverage of the Earth Liberation Front. *Crime Media Culture*, 8(1): 75–93. https://doi.org/10.1177/1741659011433366.

Joosse, Paul. 2014. Antiglobalization and Radical Environmentalism: An Exchange on Ethical Grounds. *Ethics in Progress*, 5(1): 33–51. https://doi.org/10.14746/eip.2014.1.2; https://pressto.amu.edu.pl/index.php/eip/article/view/9382/9084.

Joosse, Paul. 2021. Narratives of Rebellion. *European Journal of Criminology*, 18(5): 735–754. https://doi.org/10.1177/147737081987442.

Jordan, Rob. 2023. Low-intensity Fires Reduce Wildfire Risk by 60%. *Stanford Report*, November 10. https://news.stanford.edu/stories/2023/11/valuing-prescribed-fire.

Karlamangla, Soumya, Vik Jolly, and Jill Cowan. 2024. Wildfires Burn Dozens of Homes in California. *The New York Times* (New York edition), September 12: A21. Published online as "Dozens of Homes Burned in California as Wildfires Rage in the West" on September 11, 2024, at https://www.nytimes.com/2024/09/11/us/wildfires-california-nevada-oregon-idaho.html?smid=nytcore-ios-share&referringSource=articleShare&sgrp=c-cb&ngrp=mnp&pvid=9CD5FA26-331C-4CCE-99BD-8C29C45BFD5C.

Kateman, Brian. 2022. Why Throwing Mashed Potatoes at Monet Can Help Save the Planet. *NBC News*, October 27. https://www.nbcnews.com/think/opinion/climate-activists-throwing-food-at-paintings-could-save-the-planet-rcna54415.

Kennedy, Liam. 2021. "Whenever There's Trouble, Just Yelp for Help": Crime, Conservation, and Corporatization in Paw Patrol. *Crime Media Culture*, 17(2): 255–270. https://doi.org/10.1177/1741659020903700.

Knight, Heather, Mike Baker, and Amanda Holpuch. 2024. Fire Quickly Grows Into California's Largest This Year; Man Is Arrested. *The New York Times*, July 26 (New York edition): A11. Published online as "Park Fire Balloons in California as Blazes Spread Across the West" on July 25, 2024, at https://www.nytimes.com/2024/07/25/us/wildfires-canada-california-oregon.html?smid=nytcore-ios-share&referringSource=articleShare&sgrp=c-cb.

Komatsubara, Orika. 2021. The Role of Literary Artists in Environmental Movements: Minamata Disease and Michiko Ishimure. *International Journal for Crime, Justice and Social Democracy*, 11(1): 71–84. https://doi.org/10.5204/ijcjsd.1984.

Kudlac, Christopher S. 2007. *Public Executions: The Death Penalty and the Media*. Westport, CT: Praeger.

Lam, Anita, Nigel South, and Avi Brisman. 2025. "If There was an Observer on Mars, They Would Probably be Amazed that we have Survived this Long":

Environmental decline, Elite Escapes, and Space Colonies. In *Crime, Criminal Justice and Ethics in Outer Space: International Perspectives*, Yarin Eski and Jack Lampkin, ed. pp. 122–136. Abingdon, Oxon, UK, and New York: Routledge.

The Lancet. 2020. Australia on fire (Editorial). *The Lancet*, 395(10219) [January 18]: 165. https://doi.org/10.1016/S0140-6736(20)30098-2.

Langellier, Robert. 2024. Want to Stop Wildfire Catastrophes? Pay Firefighters More. *The New York Times* (New York edition), September 22: SR9. Published online as "We Are Running Out of Firefighters at a Perilous Time" on September 21, 2024, at https://www.nytimes.com/2024/09/21/opinion/california-wildfires-west-forest-service.html?smid=nytcore-ios-share&referringSource=articleShare&sgrp=c-cb&ngrp=mnp&pvid=C53384BF-41E9-42CE-802A-B6768FD2F835.

Lawson-Tancred, Jo. 2022a. Climate Activists Just Threw a Can of Tomato Soup on Van Gogh's 'Sunflowers' at London's National Gallery. *Artnet*, October 14. https://news.artnet.com/art-world/van-gogh-sunflowers-attacked-2192303.

Lawson-Tancred, Jo. 2022b. German Collector Hasso Plattner Has Temporarily Closed His Museum After Climate Activists Threw Mashed Potatoes at His $111 Million Monet. *Artnet*, October 24. https://news.artnet.com/art-world-archives/mashed-potato-monet-protest-2197429.

Lawson-Tancred, Jo. 2022c. German Museums Fine Climate Protestors for Damaging Property, But Environmental Groups Show No Signs of Slowing Down. *Artnet*, October 26. https://news.artnet.com/art-world-archives/german-museums-fine-climate-protestors-for-damaging-property-but-environmental-groups-show-no-signs-of-slowing-down-2199002.

Lawson-Tancred, Jo. 2022d. Guards at the Musée d'Orsay Thwarted an Attempted Soup Attack on a Vincent van Gogh Painting. *Artnet*, October 31. https://news.artnet.com/art-world/guards-at-the-musee-dorsay-thwarted-an-attempted-soup-attack-on-a-vincent-van-gogh-painting-2201533.

Lawson-Tancred, Jo. 2022e. Glass Barriers, Plainclothes Officers, Pricey Insurance: Museums Debate How to Adapt to Climate Activism in Their Halls. *Artnet*, November 7. https://news.artnet.com/art-world/security-measures-climate-protests-museums-2204699.

Lawson-Tancred, Jo. 2022f. Museum Security in Oslo Thwarted Two Activists Attempting to Glue Themselves to Edvard Munch's 'The Scream'. *Artnet*, November 11. https://news.artnet.com/art-world/oslo-munch-scream-glue-attack-2209001.

Lawson-Tancred, Jo. 2022g. Climate Protesters Hurled Maple Syrup on an Emily Carr Painting at Canada's Vancouver Art Gallery. *Artnet*, November 14. https://news.artnet.com/art-world-archives/climate-protesters-hurled-maple-syrup-on-an-emily-carr-2209586.

Lawson-Tancred, Jo. 2023. German Museums' Unexpected New Gambit to Stop Climate Activists From Attacking Their Art? Inviting Them to Drop By. *Artnet*, May 23. https://news.artnet.com/art-world/german-museums-take-a-new-tack-to-prevent-climate-activists-attacking-their-art-inviting-them-in-2307623.

Leach, Christopher. 2023. Fire at Lexington fast food restaurant forces closure, leaves extensive damage. *Lexington Herald-Leader*, September 26. https://www.kentucky.com/news/local/counties/fayette-county/article279765604.html.

Lee, Murray. 2021. Policing the Pedal Rebels: A Case Study of Environmental Activism Under COVID-19. *International Journal for Crime, Justice and Social Democracy*, 10(2): 156–168. https://doi.org/10.5204/ijcjsd.1887; https://www.crimejusticejournal.com/article/view/1887/1069.

Lennard, Natasha. 2019. How the Prosecution of Animal Rights Activists as Terrorists Foretold Today's Criminalization of Dissent. *The Intercept*, December 12. Available at: https://theintercept.com/2019/12/12/animal-people-documentary-shac-protest-terrorism/.

Lim, Clarissa-Jan. 2022. Activists Say The Outrage At Them Gluing Their Hands To Famous Paintings Is Nothing Compared To What Climate Change Will Do To Art. *BuzzFeed*, October 11. https://www.buzzfeednews.com/article/clarissajanlim/climate-activists-paintings-glue-protest.

Loadenthal, Michael. 2017. "Eco-Terrorism": An Incident-Driven History of Attack (1973–2010). *Journal for the Study of Radicalism*, 11(2) [Fa11]: 1–34. https://www.jstor.org/stable/26297171.

Lukpat, Alyssa. 2022. Protesters Arrested for Tossing Mashed Potatoes on a $110-Million Monet Painting. *The Wall Street Journal*, October 24. https://www.wsj.com/articles/climate-protesters-arrested-for-tossing-mashed-potatoes-on-claude-monet-painting-11666622214?mod=article_inline.

Lundberg, Kajsa. 2021. Visual Criminology and Lives Lived in Public Space. *City*, 25(1–2): 108–128. https://doi.org/10.1080/13604813.2021.1885915.

Lundberg, Kajsa. 2022. Moved by Fire: Green Criminology in Flux. *Crime Media Culture*, 18(1): 3–20. https://doi.org/10.1177/1741659020958450.

Lundberg, Kajsa Olivia. 2023. *Everyday Harms and Extraordinary Crises: Exploring the Edges of the City*. Ph.D. Thesis, School of Social and Political Sciences, Faculty of Arts, University of Melbourne (Victoria, Australia).

Lyman, Rick, and Gary Gately. 2005. Behind the Maryland Fires, a County in Transition. *The New York Times* (National edition), January 7: A16. Published online as "Behind Maryland Fires, a County in Transition" on January 7, 2005, at https://www.nytimes.com/2005/01/07/us/behind-maryland-fires-a-county-in-transition.html.

Mahler, Jonathan. 2008. Ahead for Obama: How to Define Terror. *The New York Times*, November 30: WK3. Published online as "Ahead for Obama: How to Define Terror" on November 29, 2008, at https://www.nytimes.com/2008/11/30/weekinreview/30mahler.html.

Malm, Andreas. 2021. *How to Blow Up a Pipeline: Learning to Fight in a World on Fire*. London and New York: Verso.

Mandler, C. 2022. Climate Activists Throw Mashed Potatoes on Monet Painting to Protest Fossil Fuel Extraction. *CBS News*, October 23. https://www.cbsnews.com/news/claude-monet-les-meules-grainstacks-mashed-potatoes-climate-activists-last-generation-germany/.

Manson, Dylan. 2024. Eco-sabotage as Defensive Activism. *Ethical Theory and Moral Practice*, 27(4) [September]: 505–522. https://doi.org/10.1007/s10677-024-10449-w.

Marchese, David. 2024. How This Climate Activist Justifies Political Violence. *The New York Times*, January 16. https://www.nytimes.com/interactive/2024/01/14/magazine/andreas-malm-interview.html.

Marshall, Alex. 2022a. Sticky Protests Hit British Museums. *The New York Times* (New York edition), July 7: C3. Published online as "U.K. Museums Face a Sticky Problem From Climate Protests" on July 6, 2022, at https://www.nytimes.com/2022/07/06/arts/design/museums-climate-protests.html.

Marshall, Alex. 2022b. Soup Thrown at Vincent van Gogh's 'Sunflowers'. *The New York Times* (New York edition), October 15: C5. Published online as "Climate Protestors Throw Soup Over van Gogh's 'Sunflowers'" on October 14, 2022, at https://www.nytimes.com/2022/10/14/arts/design/soup-van-gogh-sunflowers-climate.html.

Martin, Michael. 1990. Ecosabotage and Civil Disobedience. *Environmental Ethics*, 12(4): 291–310.

Mauney, Anna Claire. 2022. Climate Activists Glue Hands to Famous Paintings Across UK. *Art & Object*, July 5. https://www.artandobject.com/news/climate-activists-glue-hands-famous-paintings-across-uk.

Mayorquin, Orlando, Lola Fadulu, Patrick Farrell, and Joshua Barone. 2023. Climate Protesters Interrupt Met Performance of Wagner's 'Tannhäuser'. *The New York Times*, November 30. https://www.nytimes.com/2023/11/30/arts/music/met-opera-protest-tannhauser.html.

Mazurek, Jordan E. 2017. Nemo's Plight: Aquariums and Animal Abuse. In *The Palgrave International Handbook of Animal Abuse Studies*, Jennifer Maher, Piers Beirne, and Harriet Peirpoint, ed. pp. 313–336. Basingstoke, Hampshire, UK: Palgrave Macmillan.

McClanahan, Bill. 2014. Green and Grey: Water Justice, Criminalization, and Resistance. *Critical Criminology: An International Journal*, 22(3): 403–418. https://doi.org/10.1007/s10612-014-9241-8.

McClanahan, Bill. 2019. Earth-World-Planet: Rural Ecologies of Horror and Dark Green Criminology. *Theoretical Criminology*, 24(4): 633–650. https://doi.org/10.1177/1362480618819813.

McClanahan, Bill. 2021. *Visual Criminology*. Bristol, UK: Bristol University Press.

McClanahan, Bill, and Travis Linnemann. 2018. Darkness on the Edge of Town: Visual Criminology and the "Black Sites" of the Rural. *Deviant Behavior*, 38(4): 512–524. https://doi.org/10.1080/01639625.2017.1407111.

McClanahan, Bill, Avi Brisman, and Nigel South. 2017. Green Criminology, Culture, and Cinema. In *The Oxford Encyclopedia of Crime, Media, and Popular Culture*, Michelle Brown and Eamonn Carrabine, ed. New York and Oxford: Oxford University Press. https://doi.org/10.1093/acrefore/9780190264079.013.151.

McIntyre, Iain. 2021. *Environmental Blockades: Obstructive Direct Action and the History of the Environmental Movement*. Abingdon, Oxon, UK, and New York: Routledge.

McLaughlin, Eugene. 2013. Terrorism. In *The Sage Dictionary of Criminology*, 3rd edn, Eugene McLaughlin and John Muncie, ed. pp. 461–464. London and Thousand Oaks, CA: Sage.

McShane, Julianne. 2022a. Protesters Arrested After Throwing Substance From Soup Cans at Vincent van Gogh's 'Sunflowers'. *NBC News*, October 14. https://www.nbcnews.com/pop-culture/pop-culture-news/protesters-arrested-throwing-substance-soup-cans-vincent-van-goghs-sun-rcna52241.

McShane, Julianne. 2022b. German Protesters Arrested After Throwing Mashed Potatoes at Monet Painting That Sold for $110 Million. *NBC News*, October 23 [updated October 24, 2022]. https://www.nbcnews.com/pop-culture/pop-culture-news/german-protesters-arrested-throwing-mashed-potatoes-monet-painting-sol-rcna53623.

Medina, Eduardo. 2022. Monet Painting Is Splashed In Latest Stunt Over Climate. *The New York Times* (New York edition), October 25: A7. Published online as "Climate Activists Throw Mashed Potatoes on Monet Painting" on October 23, 2022 [updated November 10, 2022], at https://www.nytimes.com/2022/10/23/arts/claude-monet-mashed-potatoes-climate-activists.html.

Monbiot, George. 2022. Do We Really Care More About Van Gogh's Sunflowers Than Real Ones? *The Guardian*, October 19. https://www.theguardian.com/commentisfree/2022/oct/19/van-gogh-sunflowers-just-stop-oil-tactics.

Moore, Andrew N. Ireland. 2005. Caging Animal Advocates' Political Freedoms: The Unconstitutionality of the Animal and Ecological Terrorism Act. *Animal Law*, 11(1): 255–282. https://lawcommons.lclark.edu/alr/vol11/iss1/10.

Moore, Hilary A. 2020. *Burning Earth, Changing Europe: How the Racist Right Exploits the Climate Crisis and What We Can Do About It*. Brussels: Rosa-Luxemburg-Stiftung. Available at: https://rosalux.eu/wp-content/uploads/kontext-import/documents/4/2/3ef20f.pdf.

Morgan, Robert C. 1994. *Conceptual Art: An American Perspective*. Jefferson, NC, and London: McFarland & Company, Inc., Publishers.

Morgan, Robert C. 1996. *Art Into Ideas: Essays on Conceptual Art*. Cambridge: Cambridge University Press.

Morse, Ben. 2022. Lewis Hamilton Backs Environmental Protests, But Not Their Methods as Protestors Invade British GP track. *CNN.com*, July 4. https://www.cnn.com/2022/07/04/motorsport/lewis-hamilton-f1-protests-british-gp-spt-intl/index.html.

Morton, Adam. 2019. Yes, Australia Has Always Had Bushfires: But 2019 is Like Nothing We've Seen Before. *The Guardian*, December 24. https://www.theguardian.com/australia-news/2019/dec/25/factcheck-why-australias-monster-2019-bushfires-are-unprecedented.

Morton, Adam. 2020. Australia Records Worst December Fire Conditions After Its Hottest, Driest Year. *The Guardian*, January 8. https://www.theguardian.com/australia-news/2020/jan/09/bushfires-crisis-more-than-75-of-australia-had-worst-weather-conditions-on-record-last-month.

Natali, Lorenzo. 2010. The Big Grey Elephants in the Backyard of Huelva, Spain. In *Global Environmental Harm: Criminological Perspectives*, Rob White, ed. pp. 193–209. Cullompton, Devon, UK, and Portland, OR: Willan Publishing.

Natali, Lorenzo. 2016. *A Visual Approach for Green Criminology: Exploring the Social Perception of Environmental Harm*. London: Palgrave Macmillan/Springer Nature.

Natali, Lorenzo, and Marília de Nardin Budó. 2019. A Sensory and Visual Approach for Comprehending Environmental Victimization by Asbestos Industry in Casale Monferrato. *European Journal of Criminology*, 16(6): 708–727. https://doi.org/10.1177/1477370818788012.

Natali, Lorenzo, and Bill McClanahan. 2017. Perceiving and Communicating Environmental Contamination and Change: Towards a Green Cultural Criminology with Images. *Critical Criminology: An International Journal*, 25(2): 199–214. https://doi.org/10.1007/s10612-017-9356-9.

Natali, Lorenzo, Nigel South, Bill McClanahan, and Avi Brisman. 2022. Towards Visual and Sensory Methodologies in Green Cultural Criminology. In *Qualitative Research in Criminology: Cutting Edge Methods*, Mary Dodge and Rita Faria, ed. pp. 141–160. New York: Springer. https://doi.org/10.1007/978-3-031-18401-7_9.

News Desk. 2022a. Climate Activists Soak van Gogh's Sunflowers in Tomato Soup. *Artforum*, October 14. https://www.artforum.com/news/climate-activists-soak-van-goghs-sunflowers-in-tomato-soup-252181/.

News Desk. 2022b. Climate Activists Pelt $110 Million-Dollar Monet with Mashed Potatoes. *Artforum*, October 24. https://www.artforum.com/news/climate-activists-pelt-110-million-dollar-monet-with-mashed-potatoes-252217/.

News Desk. 2022c. Ninety-Two Museum Leaders Denounce Climate Actions Targeting Art. *Artforum*, November 10. https://www.artforum.com/news/ninety-two-museum-leaders-denounce-climate-actions-targeting-art-252255/.

News Desk. 2023. Climate Activists Attack Vélazquez Painting with Hammer at England's National Gallery. *Artforum*, November 6. https://www.artforum.com/news/climate-activists-attack-velazquez-painting-at-englands-national-gallery-519176/.

Ng, Ellie. 2023. Climate Activists Protest Against Barclays' Sponsorship Outside Wimbledon Gate. *The Standard*, July 10. https://www.standard.co.uk/news/environment/wimbledon-barclays-extinction-rebellion-climate-wandsworth-b1093367.html.

Nicas, Jack, and Thomas Fuller. 2018. Wildfire Becomes Deadliest in California History. *The New York Times*, November 12. https://www.nytimes.com/2018/11/12/us/california-fires-camp-fire.html.

Nolan, Rachael H., Matthias M. Boer, Luke Collins, Víctor Resco de Dios, Hamish Clarke, Meaghan Jenkins, Belinda Kenny and Ross A. Bradstock. 2020. Causes and consequences of eastern Australia's 2019–20 season of mega-fires (Letter to the Editor). *Global Change Biology*, 26(3) [March]: 1039–1041. https://doi.org/10.1111/gcb.14987; https://onlinelibrary.wiley.com/doi/10.1111/gcb.14987; https://onlinelibrary.wiley.com/doi/epdf/10.1111/gcb.14987.

North American Earth Liberation Front Press Office (NAELFPO). n.d. Frequently Asked Questions About the Earth Liberation Front. http://www.elfpressoffice.org/elffaqs.html (last visited September 10, 2009; on file with author; many of these "FAQs" are reproduced in Pickering (2007): 25–26, 42–64)).

North American Earth Liberation Front Press Office (NAELFPO). n.d. What is the Earth Liberation Front? http://www.elfpressoffice.org/theelf.html (last visited September 9, 2009; on file with author).

Nostrant, Rachel. 2024. Dozens of Wildfires Across More Than Half a Million Acres. *The New York Times* (New York edition), August 1: A17. Published online as "Fires Burning at 'Full Tilt' Across the Western U.S. Stretch Resources" on July 31, 2024, at https://www.nytimes.com/2024/07/31/us/wildfires-california-colorado-park-alexander.html?smid=nytcore-ios-share&referringSource=articleShare&sgrp=c-cb.

Nurik, Chloé Lynn. 2022. Facebook and the Surveillance Assemblage: Policing Black Lives Matter Activists & Suppressing Dissent. *Surveillance & Society*,

20(1): 30–46. https://ojs.library.queensu.ca/index.php/surveillance-and-society/article/view/13398/10068.

Olson, Sara. 2006. Armageddon Now!! In *Igniting a Revolution: Voices in Defense of the Earth*, Steven Best and Anthony J. Nocella, II, ed. pp. 265–269. Oakland, CA: AK Press.

Ore, Adeshola. 2022. Prized Picasso 'Unharmed' After Extinction Rebellion Activists Glue Hands to Painting in Melbourne. *The Guardian*, October 9. https://www.theguardian.com/australia-news/2022/oct/09/prized-picasso-unharmed-after-extinction-rebellion-activists-glue-hands-to-painting-in-melbourne.

Paddison, Laura. 2019/2020. 2019 Was The Year The World Burned. *HuffPost*, December 27 [updated January 2, 2020]. https://www.huffpost.com/entry/wildfires-california-amazon-indonesia-climate-change_n_5dcd3f4ee4b0d43931d01baf.

Pali, Brunilda. 2022. A Criminology of *Dis/Obedience*? *Critical Criminology: An International Journal*, 31(1): 3–16. https://doi.org/10.1007/s10612-022-09664-7.

Palumbo, Jacqui. 2022. Activists Glue Themselves to Copy of Leonardo's 'The Last Supper,' Adding to String of Similar Protests. *CNN.com*, July 6. https://www.cnn.com/style/article/just-stop-oil-protest-leonardo-da-vinci/index.html.

Parker, Jordan, and Anthony Edwards. 2024. Park Fire Rages Near Chico, evacuations ordered. *San Francisco Chronicle*, July 24. https://www.sfchronicle.com/bayarea/article/park-fire-rages-near-chico-evacuations-ordered-19595554.php.

Parsons, Andy, and Russell Goldman. 2020. Why the Fires in Australia Are So Bad. *The New York Times*, January 1 [updated January 2, 2020]. https://www.nytimes.com/2020/01/01/world/australia/fires.html.

Pawson, Chad. 2022. Climate activists throw maple syrup at Vancouver Art Gallery. *CBC News*, November 12. https://www.cbc.ca/news/canada/british-columbia/emily-carr-painting-maple-syrup-climate-activism-vancouver-1.6650085.

Pereles, Zachary. 2022. Woman attempts to glue herself to court in protest during Clippers vs. Timberwolves play-in game. *CBS Sports*, April 22. https://www.cbssports.com/nba/news/woman-attempts-to-glue-herself-to-court-in-protest-during-clippers-vs-timberwolves-play-in-game/.

Pickering, Leslie James. 2007. *The Earth Liberation Front, 1997–2002*, 2nd edn. Portland, OR: Arissa Media Group.

Polacek, Scott. 2022. Protester Chains Herself to Basket During T-Wolves vs. Grizzlies Game 1. *Bleacher Report*, April 16. https://bleacherreport.com/articles/10032905-protester-chains-herself-to-basket-during-t-wolves-vs-grizzlies-game-1.

Potter, Will. 2010. While Everyone Goes Green Uncle Sam Goes for the Throat. *Resistance, Journal of the Earth Liberation Movement*, Spring: 24–26.

Prichard, Susan J., Paul F. Hessburg, R. Keala Hagmann, Nicholas A. Povak, Solomon Z. Dobrowski, Matthew D. Hurteau, Van R. Kane, Robert E. Keane, Leda N. Kobziar, Crystal A. Kolden, Malcolm North, Sean A. Parks, Hugh D. Safford, Jens T. Stevens, Larissa L. Yocom, Derek J. Churchill, Robert W. Gray, David W. Huffman, Frank K. Lake, Pratima Khatri-Chhetri. 2021. Adapting Western North American Forests to Climate Change and Wildfires: 10 Common Questions. *Ecological Applications*, 31(8) [December]: e02433. https://doi.org/10.1002/eap.2433.

Rafter, Nicole. 2014. Introduction to Special Issue on Visual Culture and the Iconography of Crime and Punishment. *Theoretical Criminology*, 18(2): 127–133. https://doi.org/10.1177/1362480613510547.

Read, Paul, and Richard Denniss. 2020. With Costs Approaching $100 Billion, The Fires are Australia's Costliest Natural Disaster. *The Conversation*, January 16. https://theconversation.com/with-costs-approaching-100-billion-the-fires-are-australias-costliest-natural-disaster-129433.

Redmon, David. 2018. Video Methods, Green Cultural Criminology, and the Anthropocene: SANCTUARY as a Case Study. *Deviant Behavior*, 39(4): 495–511. https://doi.org/10.1080/01639625.2017.1407110.

Resistance Magazine. 2009a. Reflections on the Green Scare: What we've learned. *Resistance, Journal of the Earth Liberation Movement*, Summer: 8–11 (on file with author).

Resistance Magazine. 2009b. Earth Liberation: Movement Strategy & Direction. *Resistance, Journal of the Earth Liberation Movement*, Summer: 14–15, 17 (on file with author).

Reuters. 2022. Tour de France stage 10 halted as climate protesters dragged away by police, officials. *ESPN*, July 13. Available at: https://www.espn.com/cycling/story/_/id/34235225/tour-de-france-stage-10-halted-climate-protesters-dragged-away-police-officials.

Rosa-Aquino, Paola. 2022. Climate Protesters are Chucking Food at Priceless Paintings. A Social Movements Expert Says These Tactics Only Convince People Who Already Think Climate Change is Serious. *Business Insider*, November 19. https://www.businessinsider.com/activists-throwing-food-at-famous-art-to-protest-climate-change-2022-11#:~:text=A%20string%20of%20climate%20protests%20this%20year%20involved,concerned%20about%20climate%20change%2C%20a%20sociologist%20told%20Insider.

Rosenfeld, Ben. 2006. The Great Green Scare and the Fed's "Case" Against Rod Coronado. *CounterPunch*, March 10. https://www.counterpunch.org/2006/03/10/the-great-green-scare-and-the-fed-s-quot-case-quot-against-rod-coronado/.

Ross, Jeffrey Ian. 2007. Deconstructing the Terrorism-News Media Relationship. *Crime Media Culture*, 3(2): 215–225. https://doi.org/10.1177/1741659007078555.

Ruggiero, Vincenzo. 2017. Institutional and Anti-Institutional Violence. In *The Routledge Companion to Criminological Theory and Concepts*, Avi Brisman, Eamonn Carrabine and Nigel South, ed. pp. 311–315. Abingdon, Oxon, UK, and New York: Routledge.

Sackur, Leila. 2022. Climate Protesters Campaign by Throwing Food at Art, But Does That Work? *NBC News*, October 29. https://www.nbcnews.com/news/world/climate-protesters-campaign-throwing-food-art-work-rcna54626.

Salazar, Francisco. 2022. Climate Activists Protest Concert at Concertgebouw Amsterdam. *OperaWire*, November 4. https://operawire.com/climate-activists-protest-concert-at-concertgebouw-amsterdam/.

Salazar, Francisco. 2023. Lucerne Festival Releases Statement Regarding Climate Change Protest. *OperaWire*, September 9. https://operawire.com/lucerne-festival-releases-statement-regarding-climate-change-protest/.

Salter, Colin. 2011. Activism as Terrorism: The Green Scare, Radical Environmentalism and Governmentality. *Anarchist Developments in Cultural Studies*, No. 1(2011): Ten Years After 9/11: An Anarchist Evaluation: 211–238. https://journals.uvic.ca/index.php/adcs/article/view/17126/7340.

Schally, Jennifer L. 2018. *Legitimizing Corporate Harm: The Discourse of Contemporary Agribusiness*. Cham, CH: Palgrave Macmillan/Springer Nature.

Schneider, Tim. 2022. What the Rising Tide of Art Vandalism Tells Us About Where We Are Headed as a Culture. *Artnet*, November 8. https://news.artnet.com/market/gray-market-art-vandalism-wave-2206798.

Sciolino, Elaine, and Eric Schmitt. 2008. A Not Very Private Feud Over Terrorism. *The New York Times*, June 8: WK1, 8. https://www.nytimes.com/2008/06/08/weekinreview/08sciolino.html.

Scribner, Herb. 2022. Climate activsts arrested for throwing soup on Van Gogh. *Axios*, October 14. https://www.axios.com/2022/10/14/vincent-van-gogh-painting-soup-climate-protest.

Selig, Kate. 2024. In California, Controlled Fires Can Save Homes. Why Aren't More Happening? *The New York Times*, September 7. https://www.nytimes.com/2024/09/07/us/california-controlled-fire.html?smid=nytcore-ios-share&referringSource=articleShare&sgrp=c-cb.

Selmini, Rossella, and Anna Di Ronco. 2023. The Criminalization of Dissent and Protest. *Crime and Justice*, 52(1): 197–231. https://doi.org/10.1086/727553, https://www.journals.uchicago.edu/doi/epdf/.

Silverman, Ellie. 2023. Climate Protesters Smear Paint on Case Jousting Degas Sculpture in D.C. *The Washington Post*, April 27. https://www.washingtonpost.com/dc-md-va/2023/04/27/climate-protest-paint-degas-little-dancer-national-gallery-dc/.

Slackman, Michael. 2009. Disentangling Layers of a Loaded Term in Search of a Thread of Peace. *The New York Times*, February 26: A8. Published online as "Disentangling Layers of a Loaded Term in Search of a Thread of Peace" on February 25, 2009, at https://www.nytimes.com/2009/02/26/world/middleeast/26terror.html.

Small, Zachary. 2023. When Protesters Take Aim at Art. *The New York Times* (New York edition), July 12: C1. Published online as "After a Year of Climate Protests, the Toll Rises for Museums and Activists" on July 11, 2023, at https://www.nytimes.com/2023/07/11/arts/design/climate-protests-museums-sentences-art.html.

Smallwood, Scott. 2005. Speaking for the Animals, or the Terrorists? *The Chronicle of Higher Education*, 51(48) [August 5]: A.8.

Smith, Rebecca K. 2008. "Ecoterrorism"?: A Critical Analysis of the Vilification of Radical Environmental Activists as Terrorists. *Environmental Law*, 38(2) [Spring]: 537–576. https://law.lclark.edu/live/files/17299-38-2smith.

Smith, William. 2018. Disruptive Democracy: The Ethics of Direct Action. *Raisons politiques*, 2018/1 (69): 13–27. https://shs.cairn.info/journal-raisons-politiques-2018-1-page-13?lang=en.

Stephens-Griffin, Nathan. 2023. Biting Back: A Green-Cultural Criminology of Animal Liberation Struggle as Constructed Through Online Communiques. *Crime Media Culture*, 19(2): 252–273. https://doi.org/10.1177/17416590221110118.

Tan, Huileng. 2022. Environmental Activists Pelted a $110 Million Monet Painting with Mashed Potatoes Days After a Van Gogh Painting was Attacked by Protesters. *Business Insider*, October 24. https://www.businessinsider.com/climate-activists-monet-haystacks-painting-mashed-potatoes-environment-germany-2022-10.

Tin, Derrick, Attila J. Hertelendy, and Gregory R. Ciottone. 2021. What We Learned From the 2019–2020 Australian Bushfire Disaster: Making Counter-Terrorism Medicine a Strategic Preparedness Priority. *American Journal of Emergency Medicine*, 46(August): 742–743. https://doi.org/10.1016/j.ajem.2020.09.069; https://www.sciencedirect.com/science/article/pii/S0735675720308664#bb0020.

U.S. Census Bureau. 2000. Profile of General Demographic Characteristics: 2000. *Decennial Census, DEC Summary File 4 Demographic Profile, Table DP1.* Retrieved August 22, 2024, from https://data.census.gov/table/DECENNIALDPSF42000.DP1?q=maltby,%20washington&y=2000.

U.S. Census Bureau. 2010. TOTAL POPULATION. *Decennial Census, DEC Summary File 1, Table P1.* Retrieved August 22, 2024, from https://data.census.gov/table/DECENNIALSF12010.P1?q=maltby,%20washington&y=2010&d=DEC%20Summary%20File%201.

U.S. Census Bureau. 2020. Profile of General Population and Housing Characteristics. *Decennial Census, DEC Demographic Profile, Table DP1.* Retrieved August 22, 2024, from https://data.census.gov/table/DECENNIALDP2020.DP1?q=maltby,%20washington&y=2020&d=DEC%20Demographic%20Profile.

Vanderheiden, Steve. 2005. Eco-terrorism or Justified Resistance? Radical Environmentalism and the "War on Terror." *Politics & Society*, 33(3): 425–447. https://doi.org/10.1177/0032329205278462.

Vardoulakis, Sotiris, Guy Marks, and Michael J. Abramson. 2020. Lessons Learned from the Australian Bushfires: Climate Change, Air Pollution, and Public Health. *JAMA Internal Medicine*, 180(5): 635–636. https://doi.org/10.1001/jamainternmed.2020.0703; https://jamanetwork.com/journals/jamainternalmedicine/article-abstract/2762323.

Vegh Weis, Valeria, ed. 2021. *Criminalization of Activism: Historical, Present and Future Perspectives*. London: Routledge.

Victor, Daniel, and Maria Cramer. 2022. A Shocking Pastry Attack, but She Smiles Through It All. *The New York Times* (New York edition), May 31: A4. Published online as "Man Throws Pastry at Mona Lisa, Smearing Cream on Glass Case" on May 30, 2022, at https://www.nytimes.com/2022/05/30/world/europe/mona-lisa-pastry-louvre.html?searchResultPosition=7.

Wall, Michael. 2005. Burning Down the House. *Creative Loafing* (Atlanta), August 4–10: 25.

Wallace-Wells, David. 2024. Fires Are the Sum of Our Choices. *The New York Times*, February 14. https://www.nytimes.com/2024/02/14/climate/wildfire-climate-change-urban-firestorm.html.

Walter, Claire M., Elena K. Schneider-Futschik, Luke D. Knibbs, and Louis B. Irving. 2020. Health impacts of bushfire smoke exposure in Australia. *Respirology*, 25(5) [May]: 495–501. https://onlinelibrary.wiley.com/; https://doi.org/10.1111/resp.13798.

Watson, Angus, and Alex Stambaugh. 2022. Extinction Rebellion activists glue themselves to Picasso painting. *CNN.com*, October 9. https://www.cnn.com/style/article/picasso-extinction-rebellion-intl/index.html.

Watts, Rob. 2020. *Criminalizing Dissent: The Liberal State and the Problem of Legitimacy*. Abingdon, Oxon, UK, and New York: Routledge.

Woodhouse, Keith Makato. 2018. *The Ecocentrists: A History of Radical Environmentalism*. New York: Columbia University Press.

Wolfe, Matthew. 2022/2023. Backfire. *The New York Times Magazine*, May 29: 38. Published online as "The Rise and Fall of America's

Environmentalist Underground" on May 26, 2022 [updated June 15, 2023], at https://www.nytimes.com/2022/05/26/magazine/earth-liberation-front-joseph-mahmoud-dibee.html.

Wu, Xiao, Erik Sverdrup, Michael D. Mastrandrea, Michael W. Wara, and Stefan Wager. 2023. Low-Intensity Fires Mitigate the Risk of High-Intensity Wildfires in California's Forests. *Science Advances*, 9(45) [November 10]. eadi4123: 1–9. https://doi.org/10.1126/sciadv.adi4123; https://www.science.org/.

Yardley, William. 2008. Ecoterrorism Suspected in House Fires in Seattle. *The New York Times*, March 4. https://www.nytimes.com/2008/03/04/us/04homes.html.

Yohannes, Aron. 2022. Why did a Protester Glue Herself to the Court During the Timberwolves-Clippers Play-in Game? *The Oregonian*, April 13. https://www.oregonlive.com/nba/2022/04/why-did-a-protester-glue-herself-to-the-court-during-the-timberwolves-clippers-play-in-game.html.

Youngs, Ian. 2022. Van Gogh's Sunflowers Back on Display After oil Protestors Threw Soup on it. *BBC News*, October 14. https://www.bbc.com/news/entertainment-arts-63254878.

CHAPTER 2

Ecodefense and the Action-Communiqué Relationship

"The environmental movement is large," Farrer and Klein (2022: 218) contend, and it "encompasses a tremendous amount of variety" (2022: 220 (endnote omitted)). "There are thousands of different organizations, with millions of different members, and no end to the number of different ideas about what to prioritize and how to prioritize it," they continue (2022: 220 (endnotes omitted)). Accordingly, "[d]ifferent groups pursue their goals with different tactics," Farrer and Klein (2022: 220) explain, and these can include—and environmentalists may shift or "transition" between—"mainstream activism, protests and marches, judicial challenges, legislative and electoral politics, civil disobedience, lobbying and public relations campaigns, monkeywrenching, eco-sabotage, and FVES [forceful or violent environmental sabotage]" (2022: 220, 231 n.29 (citing Dereniowska and Matzke 2018; Johnson et al. 2010)).[1] Farrer and Klein (2022: 219) define "environmental sabotage" as "deliberate acts of property damage, aimed at preventing harm to the natural environment" (endnote omitted), and while "[s]ome of these activities are relatively minor, such as graffiti or traffic disruption," Farrer and Klein (2022: 219) state that "arson, disabling machinery, or releasing captive animals" (endnote omitted) constitute FVES.

Although "direct action" appears only in passing in Farrer and Klein (2022) and remains undefined, the term takes on more prominence

A. Brisman, *Direct Action as Conceptual Art?*, Palgrave Studies in Crime, Media and Culture,
https://doi.org/10.1007/978-3-031-76634-3_2

in other research and scholarship. Joosse (2014: 42n.10) treats "direct action" as "an umbrella term that includes acts of civil disobedience, such as sit-ins, as well as acts of sabotage and property destruction." Elsewhere in his piece, Joosse (2014: 37) mentions the "direct action tactics" of "civil disobedience" and "monkeywrenching," describing the latter as "'nonviolent resistance to the destruction of natural diversity and wilderness. It is never directed against human beings or other forms of life. It is aimed at inanimate machines and tools that are destroying life. Care is always taken to minimize any possible threat to people, including to the monkeywrenchers themselves'" (quoting Foreman and Haywood 1993: 9).

Johnston and Johnston (2017: 737) define "direct action" as "any act in defiance of the state that directly creates social change without negotiation or intermediaries (Sparrow 1997),"[2] then add that some "Radical Animal Liberation Movement" activists "take part in property destruction and so-called 'monkey wrenching' (sabotaging industry tools to prevent their use), or liberate caged animals in order to thwart animal suffering and impede the economic activities and profits of animal enterprises (Jones 2004; Liddick 2006, 2013)." The implication is that "property destruction," "monkey wrenching"/"sabotaging," and "liberating" are more zealous or fervent forms of "direct action," although they do not explicitly say so or linger on the relationship of these confrontational tactics.

Smith (2018: 13) describes "direct action" as "a mode of disruptive activism carried out with the aim of obstructing or deterring contentious practices that are opposed by activists." As Smith (2018: 13) explains, this form of action is *direct* because it "attempt[s] to stop a perceived wrong from occurring, rather than simply... attempt[ing] to communicate opposition to that wrong" (footnote omitted). While "[d]irect action can—and often does—have the effect of drawing attention to a contentious practice," Smith (2018: 13, 20) continues, "the primary rationale is not to publicize... [or] to cultivate public support or build inclusive movements." Rather, "the primary rationale" of direct action is "to *disrupt* presumed wrongdoing or perceived wrongdoers" (2018: 13 (emphasis added))—"to prevent or deter perceived forms of wrongdoing" (2018: 20).

For Smith (2018: 13), direct action is undertaken or performed as a "means of exerting pressure on perceived wrongdoers," and this can include "the use of polarizing rhetoric, threats, or the imposition of costs as a disincentive to persisting with their wrongdoing." In addition, Smith (2018: 13) points out, direct action can "function as a deterrent insofar as activists signal their intent to use it as a means of dissuading opponents from proceeding with their plans"—or, as the case may be—activists may "engage in direct action against one opponent as a means of deterring future conduct by the same and/or other opponents."

Smith (2018: 13) readily acknowledges that direct action and civil disobedience share some similarities, but he expresses consternation that more analysis and scholarship have not been devoted to how they diverge because "direct action often poses a greater normative challenge to the theory and practice of democratic citizenship." The reason, Smith (2018: 13) articulates, is that "civil disobedience is generally, albeit not universally, characterised as a *communicative* form of political protest that aims to achieve goals through moral persuasion" (emphasis added; footnote omitted). In other words, while both direct action and civil disobedience can be "means of exerting pressure," as mentioned above, the latter does so "primarily in the sense that it mobilizes the power of public opinion against inertial or intransigent public authorities" (Smith 2018: 13, 14 (footnote omitted)). "This communicative rationale," Smith (2018: 14) contends, "supports the adoption of constrained tactics on the part of civil disobedients, in that protesters have a reason to avoid behaviour—such as property destruction or violent confrontation—that might obscure their message and undermine their moral credibility in the public sphere" (footnote omitted). With direct action, however, the rationale(s) for such constraints is/are "less apparent... because activists engaged in this form of protest are much less concerned about reaching out to an audience in this fashion" (Smith 2018: 14). "Their goal," Smith (2018: 14) maintains, is "to take direct action to prevent wrongdoing, which means they might employ militant forms of conduct that are often seen as incompatible with civil disobedience as a mode of address."[3]

In contrast to Smith's (2018) attempt to articulate a distinction between "direct action" and "civil disobedience," Martin (1990: 291, 292, 293) differentiates between "ecosabotage"—"sabotage for the purpose of ecological protection" or "in the name of environmental protection"—and "legal protests concerning environmental issues." He then asks whether "ecosabotage" can be thought of or defined as "*a form* of civil disobedience" (1990: 291 (emphasis added)). Drawing on Bedau (1961), among others, Martin (1990: 296) argues that while "civil disobedience" is "illegal, done publicly, nonviolently and conscientiously with the intent to frustrate (one of) the laws, policies or decisions of *the government*" (footnote omitted; emphasis added), "ecosabotage" is "not publicly performed" and is "often aimed at frustrating the actions or policies of *private companies* and not the government" (emphasis added).[4] Moreover, Martin (1990: 296, 297) explains, while "civil disobedience" is usually undertaken with the aim of impeding one or more of the laws, policies, or decisions of a government, "ecosabotage" does not normally reflect an objection on constitutional, moral, or religious grounds *to a law*. As Martin (1990: 297) explicates, "[a]n ecosaboteur who breaks some law *L* by destroying construction equipment probably does not object on either moral or constitutional or religious grounds to the law *L* insofar as it states that it is illegal to knowingly destroy someone else's property. What he or she objects to is *the use* to which the construction equipment is put" (emphasis added). Ultimately, Martin (1990: 310, 298) concludes, "ecosabotage cannot be assimilated to civil disobedience unless one expands the definition of the latter," but that both ("ecosabotage" and "civil disobedience") can be considered "special cases of the more general concept of conscientious wrongdoing," which covers "either public *or* nonpublic law break for either religious *or* moral purposes" (emphasis in original).

Vanderheiden (2005: 427) conceptualizes "economic sabotage" or "ecotage" as "violence against inanimate objects rather than persons," and he inquires whether "radical environmentalists who commit acts of ecotage—sabotage of inanimate objects (machinery, buildings, fences) that contribute to ecological destruction—[should] be equated (by inclusion within a common definition of terrorism) with sociopaths like Timothy McVeigh who regard the slaughter of innocents as 'collateral damage' in an ideological war...." "Insofar as it does not," Vanderheiden (2005: 427) continues, "what is the nature of its moral offense, and how might it be compared with other tactics for social change?" Vanderheiden (2005: 432) posits that "acts of ecotage"—provided they are directed at specific objects that are ecologically destructive and never at living things—cannot "be condemned as the equivalent of the willful killing (or threatened) killing of persons characteristic of genuine terrorism." Thus, for Vanderheiden (2005: 432), "[e]cotage, as conceptually distinct from and less serious than terrorism, occupies an intermediate ethical space along a continuum of political tactics between terrorism (which is never justified) and civil disobedience (which may be in some cases unjustified, but is elsewhere defensible)."[5]

For purposes of this chapter and book, what is of interest is how Vanderheiden (2005) separates "civil disobedience" and "ecotage." These can be summarized in the following way. "Civil disobedience," as Vanderheiden (2005: 438, 439, 437, 434) explains: (1) possesses a rhetorical aim "that operates through the persuasion and mobilization of a political community"; (2) does not "accomplish social change by itself, but... forces deliberation or thrusts a previously ignored issue onto the public agenda"; (3) attempts to convince and "so must avoid acts which alienate its intended audience"; (4) is committed in public—"conducted in the light of day by protestors who are willing to bear legal responsibility for their actions"[6]; and (5) is "premised upon the priority of good faith negotiation, in that efforts at negotiation should first be made, and that the targeted disobedience has at its goal the return to negotiations." In contrast, Vanderheiden (2005: 438, 439) explicates, "ecotage" (1) is neither "a publicity check against injustice," nor is it "primarily a rhetorical act," and thus it "cannot serve the same watchdog role as is provided through civil disobedience"[7]; (2) is not "retaliation for some past offense but an effort to make certain present and future acts more expensive, and hence to *discourage* them" (emphasis added)—essentially, "function[ing] as a deterrent," à la Smith (2018: 13) above[8]; (3) is undertaken not with the objective of "intimidat[ing] through force or threats of violence against person, but... [to] seek change through the manipulation of the economic incentives surrounding ecologically destructive activities..... [E]cotage... aims at the balance sheet of what its perpetrators take to be an offender (or, in some cases, to physically stop ongoing acts of ecological destruction), and intentionally draw the ire of their audience"; (4) is committed "under the cover of darkness, figuratively if not literally, as individual activists aim to avoid detection while groups like the ELF [mentioned above and discussed further below] claim responsibility"[9]; and (5) is neither "civil disobedience, with its demonstrated fidelity to law and targeted legal anomaly in an otherwise just world, nor militant uprising, with its wholesale rejection of the mores of the community."

Manson (2024: 2), who leans on Bedau (1961), Martin (1990), and Vanderheiden (2005) discussed above, and Rawls (1999), noted below, centers his discussion entirely on "eco-sabotage (or ecotage)," proposing that "[s]abotage denotes two things. On the one hand, it is property damage or destruction that thwarts or obstructs some plan of action. On the other, it is to thwart or obstruct a plan of action that does not necessarily involve property damage or destruction." Focusing, even more specifically, on "eco-sabotage," Manson (2024: 2) describes it as "sabotage that involves protecting animals or the environment," admitting that this designation is "imprecise because it is capacious," but highlighting that he is more interested in ***justifications*** than ***definitions***. Manson (2024: 2) underscores that his attention is to "eco-sabotage involving property destruction because, all things being equal, it is more difficult to justify." Indeed, "[b]ecause property destruction is pro tanto impermissible," Manson (2024: 2) writes, "eco-sabotage that involves property destruction is—all things being equal—harder to justify than eco-sabotage that does not." In a footnote, Manson (2024: 2n.1) offers the following explanation:

> Although eco-sabotage is already pro tanto impermissible because it is always illegal, eco-sabotage involving property destruction is—all things being equal—worse than eco-sabotage without it. I say "all things being equal" because the act that aggregates the most components that are pro tanto wrong is not necessarily the worst since the degree of harm that results matters too. For example, I take it that illegally opening the valves of a dam to drown the inhabitants of a town is worse than illegally blowing up a dam to flood an uninhabited valley, even though both are illegal and only the latter involves property destruction.

What is crucial, then, about Manson's (2024) clarification is that "eco-sabotage" does not *always* or *necessarily* entail "property damage or destruction" (which is why I, too, as intimated above and as explained further below, do not equate the two). Indeed, as Manson (2024: 8n.12) notes, cyberattacks that result in pipeline shutdowns represent "another example of eco-sabotage that could be devastatingly effective even *without* property destruction"[10] (emphasis added). So, while "eco-sabotage" is, for Manson (2024: 2) "*always* illegal," it only "*commonly* involves property destruction" in order to "protect[] animals or the environment" (emphasis added): in other words, "eco-sabotage" may *connote* property destruction, but it does not *denote* it. Manson (2024: 2) then distinguishes between *anthropocentric eco-sabotage*, which he defines as "eco-sabotage that protects animals or the environment for the sake of humans," and *non-anthropocentric eco-sabotage*, which stands in contrast as "eco-sabotage that protects animals or the environment for their own sake." Manson (2024: 6–12) then makes his case for "eco-sabotage" as a form of defense on anthropocentric grounds, concentrating on the conditions of "just cause," "reasonable chance of success," "proportionality," "necessity," and "discrimination."

McIntyre (2021: 10), in turn, introduces the concept of "Obstructive Direct Action" ("ODA") in order "[t]o distinguish physically obstructive forms of disruptive 'direct action'" from "non-obstructive kinds of disruptive activity such as the singing of banned songs and the wearing of non-conformist clothing." For McIntyre (2021: 11), ODA encompasses "environmental blocking" ("EB"), which can occur at four different locations: (1) the "point of destruction," which "denotes locations where logging, clearing, or other development activities are threatening to directly destroy or significantly and irrevocably alter an area that has not previously been subject to such activity"; (2) the "point of production," which refers to "workplaces engaged in environmentally destructive activities"; (3) the "point of decision," which includes "summits, parliaments, corporate offices"; and (4) the "point of consumption," such as "retailers and other places from which environmentally harmful products are sold...." McIntyre's (2021) locations or "points" bear some resemblance to Resistance Magazine's (2009b: 15) notion of "effective targeting" (e.g., "producers," "retailers," "panderers"—who "create the demand for the product") and the need for ecodefenders to "take an objective look at an industry... and analyze its vulnerabilities—to look for a choke-point (where vital and core resources of the industry 'bottleneck')").

McIntyre (2021: 11) further clarifies that the tactics employed during EB can be placed into four categories: (1) "soft blockades," which "involve[] people using only their bodies, rather than devices or barricades, to occupy a site and/or block work and the passage of goods and equipment"; (2) "barricades," which "involve[] the obstruction of roads and other key points through the placement of debris, boulders, cars, and other objects"; (3) "enhanced vulnerability," whereby "activists place their bodies in devices to obstruct their opponents by both blocking and occupying a space while amplifying threats to their well-being... such as locking a body part to opponent's equipment to prevent its use or sitting in a tree to prevent it from being cut"; and (4) "sabotage," which entails "the major destruction of opponents' equipment and resources"—and which he equates with "monkeywrenching" or "pixieing."

McIntyre (2021: 10) focuses on ODA and the first three categories of EB (largely omitting examples of "sabotage," "monkeywrenching," or "pixieing"). He does, however, make note of "nonviolent direct action"

("NVDA") and adopts John Rawls' (1999: 320) definition of "civil disobedience"—the "public, nonviolent, conscientious, yet political act contrary to law usually done with the aim of bringing about a change in the law or policies of the government"—a characterization that resonates with Martin's (1990) delineation above.[11]

Brown (2021: 356) puts forth the concept of "stepwise escalation," whereby "[a]ctivists escalate from one level to the next when government and business interests prove unresponsive at the present." "[A]ctivists tend not to escalate their tactics in unison," Brown (2021: 356) explains. "Rather," he continues, "a majority continues to use legal tactics while a highly motivated minority escalates to more radical means of struggle. Even among those who escalate to radical tactics, most opt for the least confrontational options...." (2021: 356). As such, Brown (2021: 356, 357) envisions a "pyramid of environmental tactics," instead of a "ladder of escalation," in order to reflect the reality that "most activists remain at the lowest (least confrontational) level, a few climbing to the next level, and perhaps a few climbing to the next level after that until business and government accede to the activists' demands"—or, presumably, until the movement is quelled in its entirety. In Brown's (2021) "pyramid of environmental tactics," "legal tactics" (lobbying, litigation) appear at the bottom or base of the pyramid, with (nondestructive) "civil disobedience" directly above or on top of it, "sabotage" above or on top of "civil disobedience," and (interpersonal) "violence" at the top of the pyramid.

For Brown (2021: 358), "[c]ivil disobedience is the first escalatory tactic of radical environmentalism. It encompasses a variety of essentially passive techniques of resistance, such as blocking roads, locking oneself to industrial equipment, and occupying trees to prevent logging and pipeline construction through wooded areas"—activities that seem rather akin to McIntyre's (2021) notion of ODA and EB. While "civil disobedience tactics" (such as sit-down road blockades, tree sits, and maritime confrontations, discussed below) may be effective, Brown (2021: 362) contends—especially "when backed up by legal tactics that deliver victory in the form of a court ruling or legislative change"—there are other instances where the "passive resistance" of "civil disobedience" may be insufficient. In such instances, "[a]ctivists may... wish to target [an] offending corporation's physical assets, 'monkeywrenching' its destructive machinery and eating into its profit margins" (Brown 2021: 363). Thus, for Brown (2021), what distinguishes "civil disobedience" from "sabotage" is that the latter entails the destruction of property. Brown (2021:

366, 358) effectively equates "monkeywrenching" with "sabotage" and "ecotage" (or, shall I say, he uses the terms interchangeably), but what is perhaps more important is that he defines "violence"—"the highest level of tactical escalation for social movements"—as "inflicting physical harm on ***another person*** without their consent," admitting that this approach "differs from alternatives that consider the destruction of property to be violent" (emphasis added). This narrow definition of "violence" serves Brown's (2021: 358; see also 2021: 366) analytical purposes: "The word 'sabotage' clearly connotes the destruction or disabling of property without connoting harm to people. By defining violence as harm to people, not property, I ensure that the key terms—violence and sabotage—are conceptually distinct. I can thus explore the environmental movement's internal debates about the strengths and weaknesses of sabotage and violence with a minimum of confusion over terminology (Davis 1995; Pellow 2014; Scarce 2006)." Having secernated "sabotage" from "violence," Brown (2021) then turns to possible "pathways" that might lead individuals or segments of the environmental movement to escalate from "sabotage" to "violence."[12]

In this book, I use the term, "ecodefense," rather capaciously, seeing no reason, in this particular inquiry, to except civil disobedience, or to engage in a taxonomic exercise of grouping or separating "civil disobedience," "conscientious wrongdoing," "direct action," "economic sabotage" ("eco-sabotage" or "ecosabotage" or "ecotage"), "environmental sabotage," "environmental blocking" (EB), "forceful or violent environmental sabotage" (FVES), "monkeywrenching," "nonviolent direct action" (NVDA), and "obstructive direct action" (ODA). I am interested primarily in direct action that *destroys* and/or *obstructs*, as well as that which *liberates* (nonhuman animals) and, even more specifically, in the relationship of those activities to any (additional) communicative (or communicated) messages, rationales, or statements—precisely because (most commentators, including those quoted at the outset of this chapter, would concur that) the principal goal of direct action is *the action, itself*—the *destruction*, the *obstruction*, the *liberation* (or *emancipation*). The question explored in this book, then, is *the meaning* of such direct action (or ecodefense, as I have referred to it) when it—or the actions—is/are accompanied by written statements (communiqués) that may, *inter alia*, attempt to contextualize, deter (or threaten), educate, explicate, highlight (or illuminate), justify, rationalize—or connect to other crimes, harms, struggles or (presumed, perceived) wrongdoings. The means of doing so—the method of engaging in this exploration—is through a comparison or analogy, which I develop further in the next chapter.

For present purposes, we can understand "ecodefense" which, as mentioned above, includes "economic sabotage" ("ecotage" or "ecosabotage") and "monkeywrenching," as a phenomenon (or as phenomen*a*) that has (or have) occurred since at least the end of World War II, although the Boston Tea Party and the Luddite movement's attacks on mechanized looms in the early nineteenth century in England could be considered early examples of economic sabotage (see, e.g., Hays et al. 1996: 163–164; NAELFPO, n.d., *Frequently Asked Questions About the Earth Liberation Front*; Rosebraugh 2004: 20; Vanderheiden 2005: 430, 439; see generally Amster 2006: 295; Joosse 2014: 36; Lossin 2018). Ecodefense, in general, and ecotage and monkeywrenching, in particular, became popular tactics in the early 1970s and gained even more traction after the publication of Edward Abbey's ([1975] 2000) novel, *The Monkey Wrench Gang*, in which the book's four protagonists embrace and practice ecosabotage.[13] As alluded to above, examples have included firebombing, defacing, or slashing the tires of sports utility vehicles (SUVs); damaging construction equipment used for housing developments or megastores; burning buildings (such as laboratories, horse corrals, and unoccupied housing developments); disabling bulldozers and other machinery, such as seismographic equipment, used in drilling operations; removing survey stakes from road construction sites; destroying mining claim markers in wilderness areas; flattening the tires of logging trucks; blocking access to forest land that would otherwise be logged; road spiking to render logging or mining operations unprofitable; tree spiking (placing spikes in trees to fend off loggers' chainsaws); "net-ripping" (which, similar to tree spiking, involves dumping into the ocean tons of steel I-beams welded together to form large spikes that destroy bottom-trawling nets); disrupting hunts or otherwise preventing recreational hunters from hunting (known as "hunt sabotage"); crippling or otherwise dismantling research or facilities using animal-testing techniques; gluing locks at fast-food restaurants; wrecking traps for animals; and liberating or removing animals from fur farms or laboratories and industries that conduct animal-based research (see, e.g., Brisman 2008: 754–55; Brown 2020: 240; Flynn 2010; Foreman and Haywood 1993; Johnston and Johnston 2020: 567; Joosse 2012; Malm 2021: 154; Martin 1990; Shevory 1996: 187; Smith 2018; Stephens-Griffin 2023).[14]

Broadly speaking, all of these actions—and the list is certainly not exhaustive—are, as we have established, undertaken to deter, prevent, slow, or halt an environmentally destructive activity from occurring (see, e.g., Brisman 2008: 755; Hasler et al. 2020: 520; Joosse 2012: 76; Martin 1990: 292; McIntyre 2021: 13; Smith 2018: 14, 20).[15] But the *groups* dedicated to ecodefense often focus their attention on certain issues or types of environmental concern in or relating to specific areas (see, e.g., Hays et al. 1996: 166). Let us consider briefly three such groups.

The Sea Shepherd Conservation Society ("Sea Shepherd"), an international marine wildlife conservation nonprofit organization, states that its

> mission is to end the destruction of habitat and slaughter of wildlife in the world's oceans in order to conserve and protect ecosystems and species.
>
> Sea Shepherd uses innovative direct-action tactics to investigate, document, and take action when necessary to expose and confront illegal activities on the high seas. By safeguarding the biodiversity of our delicately-balanced ocean ecosystems, Sea Shepherd works to ensure their survival for future generations.[16]

In practice, Sea Shepherd pulls up long lines of poachers, engages in net ripping and other activities to impede fishing vessels, and will even use its fleet (known as "Neptune's Navy") to ram whaling, sealing, and fishing vessels on the seas—actions intended to stop violations of international environmental law, slow the extirpation of aquatic species, and address loopholes in the International Convention for the Regulation of Whaling that have permitted Iceland, Japan, and Norway to continue whaling (for scientific research and commercial purposes) despite the International Whaling Commission's (IWC) 1986 ban (see Allhoff and Buciak 2013: 221 n.5; Brisman 2008: 755 n.178–182 and accompanying text; Hale 2010: C4; Manson 2024: Martin 1990: 292; see generally Woodhouse 2018: 132–133).[17]

On land, the E.L.F., as noted above, engages in direct action, primarily in the form of economic sabotage, for the alleged purpose of stopping the exploitation and destruction of the natural environment (see Allhoff and Buciak 2013: 221n.3, 233; NAELFPO, n.d., *Frequently Asked Questions About the Earth Liberation Front*).[18] According to the North American Earth Liberation Front Press Office (NAELFPO), a news service that used to report on the covert actions taken by the E.L.F. and other individuals and groups in defense of the environment—what Brown (2020: 240) describes as "an aboveground institution that received and publicized the credit claims emanating from the ELF's cellular underground" (endnote omitted)[19]—the E.L.F.'s goals are:

(1) To cause maximum economic damage to a given entity that is profiting off the destruction of the natural environment,
(2) To educate the public on the atrocities committed against the environment and life,
(3) To make it known that any entity profiting off the destruction of life for profit may be considered the next target.[20]

Similarly, the Animal Liberation Front (the "A.L.F." or "ALF")—an anonymous, non-hierarchical, cell-structured group that predates and resembles the E.L.F. (Johnston and Johnston 2017: 737; 2020: 564, 567, 576; Loadenthal 2017: 2)[21]—"carries out direct action against animal abuse in the form of rescuing animals and causing financial loss to animal exploiters, usually through the damage and destruction of property" (A.L.F., n.d., *The ALF Credo and Guidelines*; see also Allhoff and Buciak 2013: 221 n.1). Referred to by Johnston and Johnston (2020: 563) as "the longest-standing and most notorious radical animal liberation movement worldwide (Best & Nocella 2004; Flükiger 2009)," the A.L.F., like the E.L.F., is not a centralized organization and does not have any physical membership lists (see Johnston and Johnston 2020: 564; see generally Salter 2011: 226; cf. Resistance Magazine 2009a: 9). Just as individuals or groups who carry out actions and adhere to E.L.F. guidelines may consider themselves part of the E.L.F., individuals who carry out actions according to the following A.L.F. guidelines may regard themselves as part of the A.L.F.:

1. To liberate animals from places of abuse, i.e., laboratories, factory farms, fur farms, etc., and place them in good homes where they may live out their natural lives, free from suffering.
2. To inflict economic damage to those who profit from the misery and exploitation of animals.
3. To reveal the horror and atrocities committed against animals behind locked doors, by performing non-violent direct actions and liberations.

4. To take all necessary precautions against harming any animal, human and non-human.
5. To analyze the ramifications of all proposed actions, and never apply generalizations when specific information is available.[22]

In addition to these "organizations,"[23] thousands of individuals and groups around the world have engaged in ecodefense without calling themselves "the A.L.F." or "the E.L.F."[24]

The above-quoted goals and guidelines provide some general reasons and rationales behind ecodefense. But the motivations and purpose behind *specific* actions can vary (see, e.g., Allhoff and Buciak 2013: 228, 233; Cianchi 2013, 2015; Farrer and Klein 2022: 221). And while the actions are often intended to and can, indeed, stand on their own, the reasons and rationales for specific instances of ecodefense may become clear only through a communiqué issued by the activists or an individual or entity, such as the NAELFPO, serving as a spokesperson for the activists (see Brisman 2008: 755; NAELFPO 2008). Let us now contemplate four (4) specific actions and the communiqués that accompanied them.

On October 18, 1998, fires demolished five buildings and four ski lifts at a ski resort in Vail, Colorado, causing an estimated US$12–26 million in damages, and thrusting the issue of development in wilderness areas into the national spotlight. The communiqué that followed the action stated:

> On behalf of the lynx, five buildings and four ski lifts at Vail were reduced to ashes on the night of Sunday, October 18th. Vail, Inc. is already the largest ski operation in North America and now wants to expand even further. The 12 miles of roads and 885 acres of clearcuts will ruin the last, best lynx habitat in the state. Putting profits ahead of Colorado's wildlife will not be tolerated. This action is just a warning. We will be back if this greedy corporation continues to trespass into wild and unroaded areas. For your safety and convenience, we strongly advise skiers to choose other destinations until Vail cancels its inexcusable plans for expansion.
>
> - Earth Liberation Front[25]

Apparently, the Vail expansion had been opposed by a coalition of fairly mainstream Colorado environmental groups (which included the Colorado Environmental Coalition (CEC), the Rocky Mountain Chapter of the Sierra Club, the Southern Rockies Ecosystem Project, the Biodiversity Legal Foundation, the Colorado Wildlife Federation, and Ancient Forest Rescue), as well as local residents and merchants (NAELFPO, n.d., *Frequently Asked Questions About the Earth Liberation Front*; see also Brooke 1998; Rosebraugh 2004: 60). Efforts to block the expansion—in the form of citizen comments to the United States Forest Service and suits in federal court—had preceded the Vail fires, albeit unsuccessfully (NAELFPO, n.d., *Frequently Asked Questions About the Earth Liberation Front*; see also Rosebraugh 2004: 60–71).[26]

Similarly, on March 24, 2000, in Minneapolis, Minnesota, vehicles belonging to CS McCrossan Construction were sabotaged causing US$500,000 in damages. The vehicles were being used in the rerouting of Highway 55 in Minneapolis—a construction project that was threatening the 10,000-year-old Coldwater Spring. The following communiqué was received by the NAELFPO:

> Over the weekend of March 24, maintenance work was done to the vehicles belonging to CS McCrossan construction, and were being used for the construction of the highway 55 reroute in Minneapolis. 4 machines had wires and hoses cut, dirt and sand poured into gas tanks, oil tubes and exhaust pipes, engine parts smashed, removed and destroyed, which has happened several times before and gone unreported.
>
> This site is only a quarter-mile from Coldwater Spring, a sacred site to the Dakota people, and the last source of fresh spring water in the twin cities. After cutting hundreds of old trees and decimating a once wild area, the road now threatens to cut off Coldwater's underground source, which has already happened to other springs in the city because of road construction.
>
> Also on 3 separate occasions, large amounts of rock and table salt was mixed into the sand piles being used for cement. The piles are now gone and it is assumed that they were used in the cement building the land tunnel. This is being told because salt will weaken the integrity of cement.
>
> The road is not nearly done and neither are we.
>
> -Earth Liberation Front[27]

According to Craig Rosebraugh, who served as a spokesperson for the E.L.F. from 1997 to September 2001, the Highway 55 reroute, which was threatening parkland and prairies, in addition to Coldwater Spring, had been opposed by local activists, environmentalists, and Native Americans since its beginning on August 10, 1998 (Rosebraugh 2004: 126). Opponents of the reroute had tried a variety of tactics prior to the E.L.F. attacks (including spiritual ceremonies and nonviolent civil disobedience) (Rosebraugh 2004: 126–27). As Rosebraugh (2004: 126–127) explains, "[i]n what would become another example of classic tactical escalation"—or what Brown (2021), as mentioned above, conceptualizes as "stepwise escalation"—the E.L.F. action took place only "*after* local activists had attempted and exhausted other, less severe methods" (emphasis added).

Whereas the Vail arson was an attack on the *project* (much as the Maltby arson, discussed in Chapter 1, destroyed homes in a development) and the Minneapolis sabotage was an attack on the *equipment* that would facilitate the project, in both the examples of Vail and Minneapolis, the actions occurred once *legal attempts* had failed (although those other efforts were not mentioned in the communiqués).[28] In addition, both communiqués described what transpired and explained why, although neither attempted to place the specific actions in the broader context of ecodefense or to identify a problem that extended beyond the particular environmental harm or threat. Both communiqués did, however, warn of future attacks if the projects resumed or continued.

Not all communiqués contain all of the types of information found in the Vail and Minneapolis communiqués. For instance, on January 1, 2003, the E.L.F. torched several SUVs at the Bob Ferrando Ford Lincoln Mercury Dealership in Erie, Pennsylvania, causing US$90,000 in damages. The communiqué that followed stated:

> At 5:30AM on January 1, 2003, the Earth Liberation Front attacked several SUV's at Bob Ferrando Ford Lincoln Mercury in Erie, Pennsylvania. At least four vehicles were entirely destroyed and several others sustained heavy damage, costing an estimated $90,000.
>
> Despite decades of popular environmental activism, the mainstream environmental movement, which began arguably in the early 1960s, has failed in its attempts to bring about the protection needed to stop the destruction of life on this planet. In many ways, it has served only to accelerate this destruction. It's [sic] occasional "victories", reforms or small concessions,

have fostered hope in a means of social change that has proven unable to produce tangible protection of life, time after time.

By focusing it's [sic] energy on temporary "solutions", they have altogether ignored the roots of the problem at hand. Western civilization, with it's [sic] throw away conveniences, it's [sic] status symbols, and it's [sic] unfathomable hoards of financial wealth, is unsustainable, and comes at a price. It's [sic] pathological decadence, fueled by brutality and oceans of bloodshed, is quickly devouring all life and undermining the very life support systems we all need to survive. The quality of our air, water, and soil continues to decrease as more and more life forms on the planet suffer and die as a result. we are in the midst of a global environmental crisis that adversely effects and directly threatens every human, every animal, every plant, and every other life form on the face of the Earth.

There is absolutely no excuse for any one of us, out of greed, to knowingly allow this to continue. There is a direct relationship between our irresponsible over-consumption and lust for luxury products, and the poverty and destruction of other people and the Natural world. By refusing to acknowledge this simple fact, supporting this paradigm with our excessive lifestyles, and failing to offer direct resistance, we make ourselves accomplices in the greatest crime ever committed.

Time is running out - change must come, or eventually all will be lost. A belief in state sanctioned legal means of social change is a sign of faith in the legal system of that same state. We have absolutely no faith in the legal system of the state when it comes to protecting life, as it has repeatedly shown itself to care far more for the protection of commerce and profits than for its people and the natural environment. Clearly, the state itself causes and profits from many of the various atrocities against life that we must struggle against. To place faith in that same state as though it will act in the interests of justice and life is utter foolishness and a grave mistake.

Therefore, the E.L.F. will continue to fight to remove the profit motive from the killing of the natural environment, and to draw public attention to that which is deliberately concealed from them by the forces that control our lives and destroy our home.

We urge our sisters and brothers - let us strive to become the revolutionary force we've always spoken of being, and begin to take the control of our lives out of the hands of those who would destroy us. NO COMPROMISE.

Happy New Year Bob Ferrando!

E.L.F.[29]

Like the Vail and Minneapolis communiqués, this communiqué described the action that took place. While this communiqué spelled out the E.L.F. ideology and justified ecodefense in light of the perceived failures of the mainstream environmental movement, there was little explanation as to why this particular site was selected. Whereas the Vail and Minneapolis actions represented attacks on large projects that had been opposed by many individuals using a wide variety of tactics, there was no indication as to why Bob Ferrando's dealership was chosen (as opposed to, say, another car dealership). While SUVs may, indeed, be emblematic of "irresponsible over-consumption and lust for luxury products," why did these "Elves" not torch Hummers, which are even more "irresponsible" (both in terms of fuel efficiency and danger to passengers in accidents) and more expensive (for a discussion, see Brisman 2004, 2007, 2009; Brisman and Rau 2009)? Why pick this dealership? Why at this point in time? This is not to suggest that the "Elves" *should* have torched the Hummer dealership down the road (if one existed at the time), but that the action was far more symbolic (burning four SUVs is not going to protect the environment in the way that demolishing five buildings and four ski lifts does) and that the communiqué did little more than reiterate the statement of the E.L.F. ideology that appears on the *Frequently Asked Questions About the Earth Liberation Front* page on the NAELFPO website.

A different issue arises if we examine the communiqué that followed actions to multiple targets in Georgia on December 8, 2008. The anonymous communiqué stated:

> Last week, a butcher shop in Georgia had a U-Lock placed around its security gate in the middle of the night. They were closed for two hours behind schedule the next day. This cost them hundreds in lost sales and locksmith fees.
>
> That same night, three tires on a wholesale meat company's van were slashed with a pairing [sic] knife, easily attained from the kitchen section at Safeway. This business also had their locks glued.

The next night, we cut off the gas to three fast food restaurants in the middle of a busy night.

Sabotage can be easy, risk-free, and cathartic. We have no excuses. We are the ALF.

-ALF Georgia[30]

As with the Vail, Minneapolis, and Erie communiqués, this communiqué announced what transpired. But the actions described seem the most random of all the actions considered thus far. While the implication is that the attacks were undertaken in fulfillment of the second A.L.F. guideline—"[t]o inflict economic damage to those who profit from the misery and exploitation of animals"—there was no attempt to explain why these targets were chosen or why they were attacked in these ways. Although "ALF Georgia" stated that "[s]abotage can be easy, risk-free, and cathartic"—thereby evoking the spirit of Katz's *Seductions of Crime: Moral and Sensual Attractions in Doing Evil* (1988)—we cannot discern from the communiqué why one business had its locks glued, another had its gas turned off, and a third had a U-Lock placed around its security gate.[31] In contrast to the Minneapolis attack, where damaging construction equipment was not just symbolic action but actually slowed the rerouting of Highway 55, it was not clear from the Georgia communiqué why other types of sabotage were not undertaken. Furthermore, while "ALF Georgia" asserted that they had "no excuses" for their sabotage, there was no effort on their part to situate the attacks within a broader commitment to ecodefense or A.L.F. ideology. The communiqué made no mention of a specific or even general environmental issue or problem that ALF Georgia hoped to address. And there was no indication whether future actions by ALF Georgia would occur.

Brown (2020: 237) explains that for underground leaderless resistance groups (e.g., the E.L.F.), "credit-claiming communiqués" serve both "external signaling functions"—such as "demonstrating capability, issuing demands, and soliciting public support"—and a "movement-building role."[32] As Brown (2020: 240, 237) elucidates, "autonomous cells lack direct means of speaking with one another.... [b]y issuing credit claims in 'aboveground' fora that other cells are known to read, ELF cells with no direct links are able to communicate information to one another"—in what he refers to as "the 'auto-propaganda aspect' of communiqués," which can "bolster[] morale in the militant group itself" (2020: 240

(endnote omitted)). In this way, Brown (2020: 237) continues, "[c]redit-claiming communiqués serve four critical organizing functions": (1) they can help constitute the ELF an "imagined community," as described by Anderson ([1983] 2006), of underground cells;[33] (2) they can recruit new members and thereby enable growth (because anonymous cellular groups, like the E.L.F., have no direct means of outreach);[34] (3) they can set agendas for future actions; and (4) they can "enforce[e] operational standards via an informal 'peer review' process."[35]

With the four (4) communiqués in mind (Vail, Minneapolis, Erie, and Georgia), as well as Brown's (2020) research and findings on the "external signaling functions" and "movement-building roles" (or "auto-propaganda aspects") of credit claims, we can begin to make some observations about the action-communiqué relationship and ecodefense (while bearing in mind that such a small sample cannot be representative of or generalizable to all communiqués).[36]

First, it is worth repeating that not all actions are or have been followed by elaborate communiqués. Loadenthal (2017), in his examination of 27,136 animal and earth liberation movement ("AELM") incidents, found that ninety-five percent of all attacks had been claimed through the use of communiqués, and that while "[m]any of the attack descriptions and communiqués [were] exceedingly descriptive regarding the tactics used and target selection... in some cases this vivid richness was lacking." This should be expected given the chief objective(s) of ecodefense is/are not communicative. To my knowledge, the "Elves" responsible for the Maltby fires did not transmit a communiqué following their action. The only statement to identify the fires as the work of the E.L.F., rather than an instance of random arson, was the painted bedsheet found hanging on a fence at the site. Some actions contain even less information. For instance, in April 2005, a construction site in Laguna Niguel, CA, was vandalized; only the letters "E.L.F." were left at the scene (NAELFPO, n.d., *Earth Liberation Front Diary of Actions · 2005*). Similarly, on March 7, 2005, several SUVs and trucks were spray painted with the letters E.L.F. in Fair Oaks, California. No other identifying information or explanation was offered.[37]

Second, the communiqué (when one exists) appears to serve a number of purposes. These include: (1) calling attention to/describing the action or event—something that all four communiqués did; (2) explaining the reason for the specific action—something that Vail and Minneapolis communiqués achieved; (3) illuminating the broader purpose and goals of the activists—something that the Erie communiqué stated quite clearly (see generally Stephens-Griffin 2023: 260, 265); (4) increasing awareness of environmental issues and educating the general public about a particular ecological problem (see generally Joosse 2012: 81)—something that the Vail, Minneapolis, and Erie communiqués accomplished; (5) warning that future actions will occur—something present in the Vail, Minneapolis, and Erie communiqués and suggested in the Georgia communiqué); and (6) distinguishing the action from anomic defacement or wreckage (Hays et al. 1996: 166, 175, 177; see also Bandow 1991; Rosebraugh 2004: 70; see generally Stephens-Griffin 2023: 264; Vanderheiden 2005: 432)—"shap[ing] the narrative surrounding what might otherwise be understood or perceived as apolitical or irrational acts of crime or vandalism" (Stephens-Griffin 2023: 255)—something that all the communiqués succeeded in doing, although the Georgia communiqué only by virtue of its signature associating the actions with ALF. All six (6) of these purposes possess both (or overlapping) "external signaling functions" and "movement-building roles"/"organizing functions"/"auto-propaganda aspects," in Brown's (2020) parlance. To explicate, calling attention to/describing the action or event, explaining the reason for the specific action, illuminating the broader purpose and goals of the activists, increasing awareness of environmental issues/educating the general public about a particular ecological problem, warning that future actions will occur, and distinguishing the action from anomic defacement or wreckage—all could demonstrate capability, substantiate demands/grievances/motivations, solicit public support (Brown's (2020) notion of "external signaling functions") *and* constitute the group, enable growth, set agendas, and set standards (Brown's (2020) idea of "organizing functions").[38]

Third, the nature, depth, and breadth of the communiqués may shed some light on the extent to which the activists/eco-saboteurs regard (or have regarded) the communiqué as part of ecodefense. For all four groups, the short-term goal seems to have been to raise the cost of development projects, discourage developers or animal abusers, and slow or halt specific projects. For all four groups, the long-term goal may have been more abstract—to transform society into one that is less anthropocentric and that values biodiversity over monetary wealth (see Hays et al. 1996: 168–71, 174–76; NAELFPO, n.d., *Frequently Asked Questions About the Earth Liberation Front*)[39]—but only the communiqué accompanying the Erie attack suggested such a long-term vision.[40]

Finally, while the absence of a communiqué does not necessarily indicate a general position on communiqués, one can assume that for some ecodefenders, the communiqué is—as suggested above—inconsequential or secondary to the direct action (or as indistinguishable from education—a type of indirect action). For these individuals or groups, stopping the destruction of life on the planet is all that counts. Here, "the action is the message"—"propaganda by the deed" (Joosse 2012: 80)—and there is no need to make an additional statement or release a written explanation and there may not even be a need to take credit for the action (i.e., to "sign" the action as the work of "A.L.F." or "E.L.F."). If the ecodefenders' purpose is limited to thwarting a particular project or stopping a specific environmentally destructive activity, then issuing a statement after the action is probably unnecessary; the action can stand alone. But if the activists have goals that extend beyond a particular environmental harm in a particular location at a particular time—if they see the project as representative of a larger issue or problem—and if they hope to change the way people think, feel, and act toward the environment and/or nonhuman animals, *more generally* ("mobilize the power of public opinion," in Smith's (2018: 14) parlance), then the communiqué may, indeed, become vital. In other words, if the activists' regard their ecodefense as part of a menu of tactics that can effectively help protect the environment and/or nonhuman animals (Hays et al. 1996: 174)—if the activists have not lost complete faith in the more mainstream environmental movement and still believe in the political process (although this was not the case with the Erie "Elves")—then they may be more likely to take credit for their work—to distinguish it from run-of-the-mill vandalism[41]—and, more probably, to release a statement regarding the action.

Notes

1. Earlier in their text, Farrer and Klein (2022: 218) write that "[e]nvironmental activism takes many forms. It can look conventional, like voting for the Green Party; but it can also be contentious, incorporating protests, boycotts, and... [FVES]," which they describe as "extreme"—a word that, as noted in the Preface and Acknowledgments, carries with it certain connotations.
2. At first blush, this may seem like a rather reductive definition, but the key component is "directly creates social change." As Johnston

and Johnston (2017: 747) found in their study, "[d]espite the grim prospects that animals may continue to suffer untold horrors at the hands of human beings for decades to come, RALM [Radical Animal Liberation Movement] activists are nevertheless encouraged by the passion and dedication that fellow activists express, and hold a belief that by living in line with their values and encouraging others to do the same, the world will slowly get better. Radical activists engage in prefiguration, that is to say, a collective conception of a utopic future which blossoms from the daily work and interactions of their activism" (citing Boggs 1977–1978; Haiven and Khasnabish 2014). For a meditation on animal liberation activism and prefigurative direct action and forms of resistance, see Stephens-Griffin (2023).

3. Morality is key to the argument that Smith (2018) puts forth. Smith's (2018: 14) proposition is that "there is a presumption *against* direct action" in, what he refers to as, "more-or-less democratic societies," and that this presumption is "weightier—and thus harder to overcome—than a presumption against the more constrained tactic of civil disobedience." His position, thus, is that "direct action can be a morally justified recourse for citizens, provided that it is carried out in line with what [he] characterise[s]... as an *ethic of responsibility*" (emphasis in original), which "imposes duties upon citizens to constrain the coercive and violent dimensions of their activism, while also restricting the use of direct action to the *most serious and urgent* cases of perceived wrongdoing" (Smith 2018: 14 (emphasis added)). For a comparison, see, e.g., Joosse (2014: 35), who explains that "while others have sought to assess whether the actions of the ELF can be justified morally on its own terms (Vanderheiden 2005; Brown 2007), [his] article seeks to situate radical environmentalism in a wider context of political contention, assessing the ethical and tactical feasibility ELF-style direct action"; and Amster (2006: 288, 289), who takes pains to clarify that his article is intended neither as a "moral condemnation" nor a "moral justification for the actions ostensibly taken on behalf of the environment," but that he "merely aspire[s] to generate a dialogue that touches upon both political and personal dimensions in exploring a burgeoning phenomenon that is bound up with larger issues of war and peace, security and liberty, and the relationship between humans and the biosphere."

4. Compare Martin's (1990) position with that of Lemons and Brown (2011: 3), who adopt Cohen's (1971: 111) definition of nonviolent civil disobedience as "an act of protest, deliberately unlawful, conscientiously and publicly performed. It may have as its object the laws or policies of some governmental body, or *those of some private corporate body* whose decisions have serious public consequences; but in either case the disobedient protest is almost invariably nonviolent in character" (emphasis added). (Oddly, Burkett (2016: 4), who cites to Lemons and Brown (2011) and Cohen (1971), makes reference to "protest *and* nonviolent civil disobedience (NVCD)" (emphasis added; footnote omitted)—suggesting that they are separate phenomena—but she does not explicate what kinds of NVCD would *not* constitute protest (or vice versa). She does, however, define NVCD as "the deliberate breaking of law to shift the socio-economic and legal landscape.")
5. Vanderheiden (2005: 432–33) clarifies that within the civil disobedience-ecotage-terrorism continuum, "[l]ess serious acts of resistance (and so situated between civil disobedience and ecotage) include illegal but largely symbolic acts of vandalism (e.g., billboard modification through 'subvertising') or trespass (e.g., squatting in trees slated for extraction), as these primarily aim for social change through the mobilization of public support, while ecotage instead aims primarily at the *profitability* of acts taken to be ecologically destructive" (emphasis added). "In comparison with civil disobedience," Vanderheiden (2005: 433) continues, "ecotage is plainly a more serious and objectionable tactic to employ, given its negative and galvanizing effect upon public opinion, its heightened tendencies toward disorder and lower fidelity for law, and the dangers of both its abuse and unintended consequences."
6. Vanderheiden (2005: 434) underscores that, in this way, "civil disobedience" demonstrates "fidelity to law while calling attention to a particular injustice (hence the willingness of protestors to accept legal punishment)...."
7. As such, Vanderheiden (2005: 438) adds, "the public's sense of justice cannot serve as a check upon ecotage as it does with civil disobedience (chastening protestors with silence when their complaints stray too far from public sympathies)" because "moral dissonance between protestors and the mass public is largely irrelevant to the success of the former but is fatal to the latter."

8. Essentially, Vanderheiden (2005: 438) makes clear, while "drawing public attention to ecologically destructive practices is occasionally an ancillary effect" of ecotage, "[t]he intended primary audience" of ecotage—especially actions by the E.L.F.—is "not the mass public but... the polluter or developer that is responsible for some ongoing act of ecological destruction."
9. Vanderheiden (2005: 437–438) adds that because "accepting responsibility for one's defiant acts is assumed to be an essential condition of demonstrating fidelity to the system of laws (which is itself assumed to be a necessary condition for effective political strategy), ecotage represents a more significant normative departure from normal politics than does civil disobedience. The anonymity of the saboteur may likewise work against the publicity strategy of civil disobedience, which relies for its sympathy-generating power upon media coverage of nonviolent protestors being [treated badly] by police, or otherwise by putting a human face upon an injustice."

   Vanderheiden's (2005) position resonates with that of Woodhouse (2018: 221), who writes that "[e]cotage [is]—always in symbol and at times in fact—an act of subversion. Civil disobedience [is] a means of resistance, designed to augment more conventional forms of opposition and to trigger changes in law and public opinion. One tactic sidestep[s] established institutions; the other at least in part relie[s] on them." Taking the examples of blockading, tree sitting, and tree spiking, discussed below, Woodhouse (2018: 222) elucidates how "[t]he end goal of tree spiking [is] the same as tree sitting and blockading: delaying, inconveniencing, and discouraging logging.... Blockading and tree sitting [are] civil disobedience, publicly staged and demanding recognition. Tree spiking [is] ecotage, surreptitious and anonymous."
10. In October 2016, in a coordinated action, five environmentalists in four states (Emily Johnston and Annette Klaptstein in Minnesota, Leonard Higgins in Montana, Michael Foster in North Dakota, and Ken Ward in Washington) accessed and turned the emergency shut-off valves on four pipelines carrying oil into the United States from Canada (see, e.g., Federman 2020; Holthaus 2019; Kuipers 2018; Moran 2018; Nicholson 2019; Nijhuis 2018; Puckett 2020; Stephenson 2018). According to Manson (2024: 8), the "Valve

Turners," as they came to be known, "did not destroy property (apart from the chains that secured the valves) to achieve their goals." As such, Manson (2024: 8) declares, "[e]ffective eco-sabotage need not involve property destruction." Indeed, "[t]he most materially costly actions," Manson (2024: 8) asserts, "and the most effective in environmental defense, need not involve property destruction." For more on other efforts to shut down valves on oil pipelines, see, e.g., Associated Press (2019), Gray-Donald (2015a, 2015b).

11. Like McIntyre (2021), Manson (2024), discussed above, also finds Rawls (1999) helpful. According to Rawls (1999), Manson (2024: 3) claims,

    > civil disobedience occupies a space between conscientious objection and revolt. The former is illegality based on one's private morality and the latter is an outright rejection of government. Civil disobedients do neither of these things and instead appeal to a public sense of justice by, for example, publicly announcing their intentions and willingly submitting to arrest, they draw attention to the injustice of the law they reject. For Rawls and others, civil disobedience is an illegal, yet public, nonviolent, and conscientious act done with the aim of changing the law (Rawls 1999, §§ 55, 57, 59; Bedau 1961) [footnote omitted].

    Manson (2024: 3) points out, as others do (see Chapter 1), that "eco-sabotage" is "often clandestine, not public." "In many cases," Manson (2024: 3) continues, "eco-saboteurs would thwart their own aims by declaring their acts in advance. So, if permissible eco-sabotage is illegal action that appeals to the public sense of justice, it seems like civil disobedience but undermines the Rawlsian definition."

12. "The first pathway," Brown (2021: 366) describes, is "one of frustration and necessity"—what he refers to as "The Pathway of Necessity." "[I]f environmentalists' legal activism, civil disobedience, and sabotage fail to achieve their goals," Brown (2021: 366) explains, "some activists may view violence as a logical next step, a tactic of last resort when less radical options have proved ineffective." "The second pathway to violence," Brown (2021: 366) continues, "is through the escalation of environmentalists' aims: if activists set their sights beyond environmental protection

to *political revolution*, some are likely to escalate their tactics to challenge the state's monopoly on the use of force" (emphasis in original). What is interesting about this second pathway—which Brown (2021: 368) calls "The Pathway of Revolution"—is that environmentalists "may view violence as necessary to overthrow the political system" if—or when—they "cease to think of themselves as single-issue activists and come to see themselves as part of a broader revolutionary movement . . . ."

13. For a discussion of the role of *The Monkey Wrench Gang* in influencing Earth First!, see Allhoff and Buciak (2013: 229–30); Brown (2021: 363); Hays et al. (1996: 164); Joosse (2014: 37n.5 and accompanying text); McIntyre (2021: 80); Martin (1990: 292).
14. It bears mention that ecodefenders may employ more than one of these tactics or techniques in a given action or event (see, e.g., NAALFPO 2008). For example, Stephens-Griffin (2023: 259), in his research on animal liberation-related direct action in 2020, found that, for example, "vandalism often occurred alongside sabotage." By the same token, some (groups of) ecodefenders may explicitly reject one or more than one of these tactics or techniques. For example, Brown (2021: 364–365), Joosse (2014: 37), Vanderheiden (2005: 440–441), and Woodhouse (2018: 1–2, 221–228) all discuss how and why Earth First! came to renounce the tactic of tree spiking. Similarly, albeit on a more individual level, Johnston and Johnston (2020: 568) describe how a formerly incarcerated A.L.F. activist "frame[d] the actual liberation of animals as the most direct and effective way of achieving justice, an idea which is widely shared by fellow activists within the [Radical Animal Liberation Movement] (Liddick 2006)."

    Note that certain tactics or techniques of activism/ecodefense may be more common in particular places at specific times than others. For instance, Stephens-Griffin (2023: 258) points out that "fox hunt sabotage" is "a vibrant and common form of animal activism in the UK (Cox and Donovan 2020)," but that, oddly, his dataset (a sample of 134 communiqués published on *Bite Back*'s website), included no such examples. (For accounts of hunt sabotage, see Chuk'shown Earth First! (2004); Curnow-Garland (2010); see generally Loadenthal (2017:1).)

    Note also that some forms of activism/ecodefense may comprise or necessitate more individuals/participants than others, but that some seemingly more "solitary" tactics actually involve (much) larger groups of people (see, e.g., Fountain 2006; Hill 2000; Oak 2022; see generally Martin 1998; Oldenburg 2004). On the differences between "fur farm raids" and "live liberations, such as lab raids," in terms of the number of "volunteers" that may be required, see, e.g., Resistance Magazine (2009a: 9).

15. According to Hays and colleagues (1996: 165, 174), "most of these [actions] are undertaken to directly halt or at least slow development projects.... For some radical environmentalists, sabotage is an act of desperation, an effort to lash out at a system that they see as destructive and beyond their control (Mills 1991: 167)." Similarly, Farrer and Klein (2022: 221) remark that "FVES is more likely when conventional tactics have been tried and floundered" (citing Farrer and Klein 2017). Smith's (2018: 17) take is a little different: "direct action campaigns do not arise in a vacuum, but are almost always carried out by groups or networks of activists that have entered into a morally significant relationship with a broader constituency of like-minded others"—a relationship that he explores further in the rest of his article and which I comment on elsewhere in this book. Finally, Hays and colleagues (1996: 174) suggest that some activists may not fixate on their relationship to other environmentalists and their politics and tactics, and may, instead, "view sabotage as a way of reclaiming their identity as natural, spiritual beings in an increasingly artificial world (Foreman 1991: 115–146)."

    For an in-depth examination of the "radical environmental activists'" reasons for and meaning of direct action campaigns to protect whales in the Southern Ocean and to prevent clear-felling of old growth forests in Tasmania, Australia, see Cianchi (2013, 2015). For an analysis of three possible existential motivations for engaging in terrorism—the desire for excitement, the desire for ultimate meaning and the desire for glory—see Cottee and Hayward (2011). While Cottee and Hayward (2011: 965) argue that "terrorism, for those who practice and embrace it, can be profoundly thrilling, empowering and spiritually intoxication," they make clear that "no single explanatory perspective can possibly do full justice to a phenomenon as complex and diverse as terrorism" (2011: 981 n.17)—and I believe that is true for ecodefense as well. In addition, Cottee and Hayward (2011: 966) emphasize that in identifying and outlining existential motives for terrorism, their purpose "is not to formulate a general theory of terrorist motivation." By the same token, this book seeks neither to promulgate a general theory of ecodefense motivation nor a general theory regarding the use of communiqués.

16. Sea Shepherd Conservation Society, n.d., *Who We Are: Sea Shepherd's Mission Statement*.
17. Smith (2018: 23) reflects on "circumstances where a suitably authoritative process has culminated in a decision to regulate, restrict or prohibit a particular practice, but recalcitrant parties either refuse to accept the outcome or exploit loopholes in violation of its spirit." In such situations, Smith (2018: 23) contends, "[d]irect action aimed at deterring or preventing this recalcitrant behaviour may be defended by activists, through appealing to the moral authority of the decision-making process." In such cases, Smith (2018: 23) continues, "[d]irect action may... be seen as more appropriate than civil disobedience... as an authoritative process has already run its course and the immediate priority is not to instigate further reflection but to ensure that the original decision is enforced." As an example, Smith (2018: 23–24) points to "the use of direct action tactics by activists at sea to disrupt or prevent whaling." Smith (2018: 24) explains:

> The relevant context is the imposition of a moratorium on whaling by the International Whaling Commission (IWC), with exceptions that allowed Norway, Iceland and Japan to engage in "scientific" whaling. The claim of anti-whaling activists, and others, is that "scientific whaling is actually a cover for a continued (but sharply reduced) commercial catch; the whale products are still used as before, and precious little science comes from the dead whales". The anti-whaling activists thus frame their direct action not as a communicative protest, but as an informal attempt to police the moratorium. (quoting Dryzek 2000: 126)

Indeed, because the international whaling regime does not possess effective enforcement mechanisms—a sad truth endemic to many transnational governmental treaties or arrangements—many anti-whaling ecodefenders view their activities as attempting to *enforce* and *uphold* (the) law, rather than as illegal.

18. Cf. Brown (2021: 365), who states that "[t]he ELF's strategic logic is *not* to blockade businesses to slow their predatory activities, but to *punish them* via property destruction and economic attrition."
19. Brown (2020: 241) fleshes out the nature of the relationship between the NAELFPO and the E.L.F., articulating how, "[d]uring the period of its most intense activity, the North American branch of the ELF developed a hybrid underground/aboveground organizational structure," whereby [u]nderground cells carried out sabotage actions in anonymity," and then, "[u]pon completing their acts, cells would issue credit claims to the [NAELFPO], which operated independently aboveground, with a mailing address, telephone number, and email account." The NAELFPO would then "publish[] the text of credit-claiming communiqués online and in press releases to mainstream media and radical publications...." (Brown 2020: 241). For more on the relationship of the NAELFPO to the E.L.F., see, e.g., Joosse (2012: 81; 2021: 739–740).
20. NAELFPO, n.d., *Frequently Asked Questions About the Earth Liberation Front*; see also Allhoff and Buciak (2013: 233).

    The E.L.F.'s goals closely resemble its guidelines (see Pickering 2007: 53). The first two are identical, but the third guideline encourages individuals and groups: "To take all necessary precautions against harming life." As NAELFPO explains, "[t]he guidelines are very similar to the goals of the ELF with one major exception. The third of three guidelines states that one must take all necessary precautions against harming life. The ELF considers itself a nonviolent organization as no physical harm has come to a human as a result of the group's actions. This is by no means a sheer coincidence but in fact a commitment to the guidelines" (NAELFPO, n.d., *Frequently Asked Questions About the Earth Liberation Front*; see also Allhoff and Buciak 2013: 233; Dickel 2012: 255n.4 and accompanying text; Rosebraugh 2004: 18; see generally Brown 2020: 240, 246, 247; Farrer and Klein 2022: 219; Martin 1990: 295, 301; Smallwood 2005).

    Leslie James Pickering, who co-formed the North American Earth Liberation Front Press Office (NAELFPO) with Craig Rosebraugh, reproduces the "ELF Guidelines" in his book, *The Earth Liberation Front, 1997–2002* (2007: ix, 53; see also Brown 2020:

241), albeit with slightly different wording than those quoted in the body of the text:

> 1. To cause as much economic damage as possible to a given entity that is profiting off the destruction of the natural environment and life for selfish greed and profit.
>
> 2. To educate the public on the atrocities committed against the environment and life.
>
> 3. To take all necessary precautions against harming life.

I subscribed to the first four (print) issues of *Resistance, Journal of the Earth Liberation Front* ("Resistance Magazine"); three of the first four issues (Resistance Magazine (Summer 2009: 33 (endpaper)), (Fall 2009: 47), (Spring 2010: 56)) published yet a third version of the "ELF Guidelines":

> 1. To cause maximum economic damage to a corporation, government agency or other entity that is profiting from the destruction of the Earth.
>
> 2. To reveal and educate the public on the threats to the Earth and the necessity of direct action to stop them.
>
> 3. To take all necessary precautions against harming human or animal life.

The point is to highlight the unifying spirit and principles of the E.L.F., not to try to expose some sort of inconsistency with respect to language. Those who undertake ecodefense and identify with the E.L.F. in the process carrying out their actions, share a commitment to protecting the Earth, not to specific phrasing.

21. Loadenthal (2017: 2) claims that the A.L.F. was founded in England in 1976, but that "[i]n the United States, the origin story of the ALF is a bit murky...." Regardless of when the A.L.F. formed in the United States, Loadenthal (2017: 2) asserts that by 1994, it had inspired "the formation of an organizationally and tactically similar movement targeting institutions of *ecological* exploitation through methods of sabotage and vandalism—the Earth Liberation Front (ELF)" (emphasis in original). While Farrer and Klein (2022: 220) observe that the A.L.F. and E.L.F. "operate through individual cells with no explicit links to each other or to other environmental organizations" (endnote omitted), Brown (2020: 249,

245) refers to the A.L.F. as the E.L.F.'s "sister group" and notes instances of "fur farm liberations" in Wisconsin and Minnesota that were "claimed jointly" by the A.L.F. and the E.L.F. Given the shared goals of both the A.L.F. and the E.L.F., as well as the desire and need to act in a clandestine fashion, it is probably safe to state that the A.L.F. and the E.L.F. have worked in conjunction with each other (or, shall I say, individuals who have identified with the A.L.F. may have collaborated with those who have performed actions in the name of "the E.L.F." and vice versa)—(perhaps) without even knowing so. In his study, Loadenthal (2017: 2) effectively combines the A.L.F. and the E.L.F., referring to the "animal and earth liberation movements" as "AELM." Johnston and Johnston (2017: 738) describe the E.L.F. as "modelled" on "its sister organization"—the A.L.F.—but instead of merging them in the way that Loadenthal (2017) does, they adopt the name "Radical Animal Liberation Movement" (or "RALM" or see note 2 above). In their studies, Johnston and Johnston (2017, 2020) include the A.L.F. in the RALM, as well as the Frente de Liberación Animal (FLA) in Mexico, and the Earthling Liberation Kollective (ELK) and Grassroots Ontario Animal Liberation (GOAL) in Canada.

22. A.L.F., n.d., *The ALF Credo and Guidelines.*

Johnston and Johnston (2017: 748n.6; 2020: 577n.1) list the "ALF Guidelines" as follows:

> (i) To liberate animals from places of abuse... and place them in good homes where they may live out their natural lives, free from suffering.
>
> (ii) To inflict economic damage on those who profit from the misery and exploitation of animals.
>
> (iii) To reveal the horror and atrocities committed against animals behind locked doors, by performing direct actions and liberations.
>
> (iv) To take all necessary precautions against harming any animal, human and non-human.
>
> (v) Any group of people who are vegetarians or vegan and who carry out actions according to these guidelines have the right to regard themselves as part of the Animal Liberation Front.

As with the "ELF Guidelines," the "ALF Guidelines" reprinted in Johnston and Johnston (2020: 577n.1) contain slightly different phrasing than those quoted in the body of the text of this chapter.

Again, the point is not to try to expose some sort of inconsistency with respect to language, but to accentuate the unifying spirit and principles of the A.L.F. Those who undertake ecodefense and identify with the A.L.F. in the process carrying out their actions, share a commitment to ending nonhuman animal exploitation, suffering, and use, not to specific wording.

23. Although Pickering (2007: 42) describes the E.L.F. as "an international underground organization that uses direct action in the form of economic sabotage to stop the exploitation and destruction of the natural environment" (see also Brown 2020: 241), I (still) put "organizations" in quotation marks because, as mentioned in the body of the text above, the E.L.F. and A.L.F. do not revolve around a physical base or headquarters with a board, officers, staff, and personnel, nor do they hold regular meetings and fundraising campaigns. With respect to the E.L.F., Pickering (2007: 52) explains that "[t]he ELF does not have any sort of physical membership list or meetings you can attend to become involved." "Remember," he writes, "the ELF revolves around not a physical base or classically-designed structure [and here is where the ALF and ELF differ from Sea Shepherd (see Sea Shepherd Conservation Society, n.d., *Crewing at Sea FAQ*)], but instead an ideology. If you believe in the ELF ideology and you follow a certain set of widely-published guidelines [see previous note and accompanying text], you can conduct actions and become part of the ELF" (2007: 52; see also Brown 2020: 241). Correspondingly, Stephens-Griffin (2023: 253) states at the outset of his article that, "[a]s opposed to resembling a traditional group with established leadership positions and hierarchies, the ALF represents a form of leaderless resistance. Within this, independent, autonomous cells and individuals engage in action unilaterally, providing it adheres to the key principles of the movement" (endnote omitted). Analogously, Johnston and Johnston (2020: 564) make plain that "the ALF is not a group or organization in the traditional sense. The ALF is the name used by the global network of autonomous and anonymous cells of clandestine animal liberation activists who follow ALF guidelines and engage in direct action against animal enterprises, such as freeing animals from fur farms, or damaging property and tools used to contain or kill animals." In a similar vein, Joosse (2012: 78) makes clear in his decision to refer to "the

ELF" in the singular: "I do not intend to convey a sense that the ELF is characterized by significant levels of organization unity. Rather than a 'group' or an 'organization,' the ELF is a collectivity in the most limited and virtual sense (Joosse 2007)." The E.L.F. and A.L.F. are not alone in this regard. As noted above, the E.L.F. is an offshoot of Earth First!—itself a group without formal, physical membership lists. Indeed, the Earth First! website declares, "Earth First! is not an organization, but a movement. There are no 'members' of Earth First!, only Earth First!ers" (Earth First! n.d.).

24. For example, in July 2009, a group called "Combatientes Ecologistas por la liberacion de la tierra" (or "Environmentalist Combatants for the Liberation of the Earth") engaged in an act of ecodefense and sent a communiqué to the North American Animal Liberation Press Office (NAALFPO 2009). In a different vein, in January 2008, Agence France-Presse (AFP 2008) reported that a vessel belonging to Greenpeace, an organization with membership lists, had trailed Japanese whalers in Antarctic waters in an effort to prevent them from killing giant sea creatures—direct action undertaken in conjunction with the more militant Sea Shepherd—which Brown (2021: 363) refers to as "a more radical offshoot of Greenpeace." (Sea Shepherd's founder, Paul Watson, was (also) a founding member of Greenpeace, as well as a field correspondent for Defenders of Wildlife (from 1976 to 1980), and a field representative for the Fund for Animals (1978 to 1981); he served on the board of directors of the Sierra Club from 2003–2006.)
25. NAELFPO, n.d., *Frequently Asked Questions About the Earth Liberation Front*; see also Pickering (2007: 13, 48); Rosebraugh (2004: 60); see generally Brown (2020: 242).
26. To underscore the E.LF.'s efforts to take all necessary precautions against harming life, Wolfe (2022/2023) reports that when the "Elves" set fire to the resort's buildings and ski lifts, they skipped one of the structures after peering inside and discovering two sleeping hunters.

    The action in Vail has been the subject of much discussion and commentary. For an in-depth examination, see Glick (2003); for shorter notes and observations, see, e.g., Amster (2006: 288, 297); Brown (2020: 241, 242); Combs (2006: 180); Joosse (2012: 78, 83, 85; 2014: 39; 2021: 739, 743); Loadenthal (2017: 23); Vanderheiden (2005: 426).

27. NAELFPO (2000); see also Rosebraugh (2004: 126).
28. For another example of an E.L.F. action transpiring after more conventional opposition had failed, see Associated Press (2009); NAELFPO (2009). For a more general discussion of the extent to which legal means should or might be exhausted prior to illegal ones, see Martin (1990: 299–300, 304–09). While it might be tempting to think of a *progression* from legal to illegal tactics—a "stepwise escalation of intensity" (Brown 2021: 356)—in circumstances where multiple groups/organizations and social movements are involved in the same environmental campaign or struggle, each may, in fact, "contribut[e] according to its tactical comparative advantage" (2021: 361). As Brown (2021: 362) notes, "legal tactics take time to work. It may fall to a small contingent of committed civil disobedience activists to grind an environmentally destructive project to a halt and hold the space while legal actions wind their way through the courts." Thus, Brown (2021: 362) points out, "[t]he successes of key environmentalist campaigns cannot be traced to a single tactic or organization, but to a *synergistic combination* of legal action and radical civil disobedience by diverse coalitions of activists and groups" (emphasis added). This would seem to resonate with Burkett's (2016: 43) contention that "[e]mploying both 'insider' and 'outsider' tactics can achieve significant policy gains" insofar as "[t]he larger the movement infrastructure, the greater the ability to adopt 'multiple influence methods.' In short, 'neither disruptive protest nor institutional influence activity alone is sufficient to make favorable changes in policy,' militating in favor of mending fences...." (quoting Johnson et al. 2010: 2271). For Taylor (2013), "direct action resistance can bring attention to issues in a way that electoral politics and lobbying cannot. It can inspire action and apply political pressure on corporate and government officials." That being said, Taylor (2013) cautions that "the more an action risks or intends to hurt people, the more the media and public focus on *the tactics* rather than the concerns that gave rise to the actions. This means that the most radical tactics tend to be counterproductive to the goal of increasing awareness and concern in the general public" (emphasis added). Brown (2021: 364), however, calls Taylor's (2013) argument into question, pointing to a study by Farrer and Klein (2022: 218) that, in an analysis of electoral

politics in the United States, found that environmentally motivated sabotage can have "a negative impact on voter attitudes toward *all* environmental organizations," but that this effect can be rendered void if voters believe that "democratic methods ha[ve] failed" to achieve the necessary reforms (emphasis added). Because "sabotage tends to be an *escalatory* tactic rather than a tactic of first resort," Brown (2021: 364) suggests, "its negative effect on environmentalists' public image is likely to be mitigated in many cases by a history of failed legal activism that demonstrated the need for more radical steps" (emphasis added). The point that I wish to make in this book is that the extent to which an act of ecodefense alienates potential supporters will likely depend on a number of factors (e.g., the nature of the cause, the actual and/or perceived immediacy of the harm, the efficacy of other strategies and tactics, the degree of state repression)—one of which is the presence of and nature of a communiqué to the act of ecodefense.

29. NAELFPO (2003).
30. NAALFPO (2008).
31. Echoing Cottee and Hayward's (2011: 965) observations about terrorism (see n.15 above), Smith (2018: 13) underscores that direct action is "sometimes defended as a particularly empowering form of disruption, as it emphasizes taking action to achieve desired change rather than merely calling upon others to help bring that change about" (footnote omitted). Relatedly, the "Editor's Note" (2009c: 5) of the second edition of *Resistance, Journal of the Earth Liberation Front* attests to the emotional dimensions of ecodefense: "A glued lock, a smashed window, or spray-painted wall can be a visceral and legitimate expression of frustration at the impotence many of us feel when faced with seemingly insurmountable odds and an opposition with the world's wealthy elite and law enforcement as stakeholders in the status quo." Similarly, Joosse (2021: 741) posits that "though its messaging the ELF harmonizes individual and collective interests by providing a relatively safe opportunity for emotionally satisfying action (whether through the experience of excitement, fun, or through the discharge of righteous anger." "Specifically," Joosse (2021: 741) continues, "it conveys the sense that those who adopt the guise of an 'elf' can strike enemies and enjoy a temporary empowering reversal of the asymmetries that usually ensure the dominance of 'mere creatures'

by 'mighty corporations.' Such 'sneaky thrills' are well known to deviance theorists (Katz 1988)...."

32. Brown (2020: 239–240) describes how "credit claims carry external signaling value" insofar as they "attribute[e] the destructive capability demonstrated in an attack to the group that carried it out" (endnote omitted). In addition, Brown (2020: 240) continues, "[c]redit claims... give the group an opportunity to explain its grievances and demands," and that "[t]his propaganda content is directed at external audiences: the government; other militant groups with which the perpetrator claims solidarity; or the broader ethnic, religious, or political constituency the group claims to represent" (endnotes omitted). While Brown (2020: 239, 242) focuses not on how communication may be "directed... 'outward' to nonmembers of [a] militant group," but on the ways in which "credit-claiming communiqués" may be "written for [a] movement," he acknowledges that "[t]he ELF's credit claims served some of the same external signaling functions emphasized in existing terrorism literature," such as "to explain the group's motivation and demands, making the organization appear more moderate in the eyes of the public" (endnote omitted) and "to educate the ELF's targets as well." In this way, Brown (2020: 249) maintains, the E.L.F.'s credit-claiming communiqués could be characterized as both "[n]otes from the underground" and "notes *to* the underground" (emphasis in original).

    The flavor of Brown's (2020) approach appears in Joosse (2021), although the latter author does not make explicit reference to the former's distinction between "external signaling functions" and "movement-building roles." Joosse (2021: 740, 742) mentions how communiqués can help spread the E.L.F.'s messages and issue warnings to potential targets, as well as "lend individual actions a form of ***tactical legitimacy***" (emphasis in original)—which are essentially "external signaling functions"—and "promulgate[] a folk theory of ***solidarity*** that depicts individual participants as nesting within a larger imagined community"—which is a "movement-building role" (emphasis in original). (Indeed, Brown (2020: 240; see also 2020: 249) suggests that communiqués can help "constitute the ELF itself as an imagined community of underground cells"—a "critical organizing function.")

33. This feature can prove vital for leaderless underground organizations, Brown (2020: 243) contends, because "[w]ith no direct links and no leadership, no single person can establish that a 'group' exists, determine who is a member, or demonstrate that any action relates to any other."
34. Brown (2020: 245) adds that "[t]he ELF's only path to growth is to inspire copycats. To this end, ELF communiqués seek to convey a sense of belonging—to an emotional community one may join by taking on the group's revolutionary mission."
35. This is particularly important because "[t]he guidelines for ELF actions [a]re stated clearly in print," which meant that when the NAELFPO was in operation, "[c]ells could use the forum of the press office to comment on and affirm one another's good work—or reject bad work" (Brown 2020: 247). Essentially, Brown (2020: 247) claims, this threat of rejection served to reinforce (the ELF's) behavioral standards, "creat[ing]... negative social pressure to uphold the ELF's noninjury norm" which, he maintains, is arguably "[t]he most important standard" (2020: 246). So effective was this "peer review process of sorts," Brown (2020: 247) points out, that the NAELFPO "never had to disown any action claimed on behalf of the ELF," which (as noted in Chapter 1 and Chapter 3) "has yet to harm a human victim" (endnote omitted).

    Interestingly, while the E.L.F. (and A.L.F.) subscribe to a "noninjury norm" and take precautions against harming human or animal life—and the "peer review process" upholds this behavioral standard—Islamic militants, who *do* kill people, also follow a sort of list of rules for guiding and justifying taking a life. This kind of "jihadi etiquette" (or set of "jihadi tenets") is not written down—unlike the "ELF Guidelines" and the "ALF Guidelines"—but they do have "deep roots in the Middle East" (Moss and Mekhennet 2007).
36. While it might be misleading to speak of a "typical" *communiqué*, one can identify *patterns* in the A.L.F.'s and the E.L.F.'s methods of ecodefense. For example, in his study of more than 27,100 AELM "attack incidents," Loadenthal (2017) describes trends with respect to: "group claim" (i.e., chosen names or monikers included in claims of responsibility); tactics employed (e.g., vandalism, sabotage, fire, animal liberation, "antipersonnel tactics"); targets selected (based on factors such as accessibility,

perceived vulnerability, relationship to an ongoing campaign, and the symbolic and communicative result of subsequent publicity); and the relationship of tactical selection to target choice. Thus, it may be easier to speak of an "average act of eco-terrorism," to use Loadenthal's (2017: 15) than an average communiqué (about average acts).

37. Resistance Magazine (2009a: 11) elucidated in its premiere issue at least one reason for not issuing a communiqué:

> Communiques have typically been used to publicly attribute a political message or explanation to an anonymous direct action. During sentencing phase . . . communiques for specific actions [have been] used against the accused. These communiques [have been] used to apply terrorism enhancements which, in many cases, [have] resulted in longer prison sentences and justification for harsher treatment/restrictions once in side the federal prison system.
>
> This begs the question as to whether the benefit/reward of such communiques outweighs the risk that they may somehow provide forensic or identifying information about the author in addition to likely being used against her/him in future court proceedings. Are a few sentences worth a few years in prison?

38. Even the fifth purpose—"warning that future actions will occur"—which could be considered a *deterrent* function or aspiration (see Smith 2018: 13, 20; see generally Vanderheiden 2005: 432, 438, 441, 443, 447n.24) and thus seemingly more akin to an "external signaling function"—could (also) serve to encourage/recruit new cells (by suggesting growth) and "solve... agenda-setting problems" (Brown 2020: 246).
39. Such an ideology is consistent with the principles of "deep ecology," the term and ecophilosophy developed by Arne Næss (1973, 1984, 1986), although one need not engage in ecodefense in order to be a "deep ecologist." On the relationship of Earth First!, A.L.F., and E.L.F. to "deep ecology," see Malm (2021: 153–157); see also Brown (2020: 243; 2021: 369); Farrer and Klein (2022: 221); Joosse (2014: 36n.3 and accompanying text); for an in-depth discussion of "deep ecology," see, e.g., Devall and Sessions (1985); Drengson and Inoue (1995). Oddly, Woodhouse (2018: 2, 293n.2, 96) equates "deep ecology" with "biocentrism"

and "ecocentrism," claiming that "radical environmentalists used the terms [biocentrism and ecocentrism] interchangeably," and stating that "[t]he philosophical foundation for 1980s and 1990s environmentalism had many labels. Generically it was 'ecocentrism' or 'biocentrism,' but its brand name, by the 1970s, was 'deep ecology.' Whatever its name...." Woodhouse's (2018) inattention to the philosophical differences between "biocentrism" and "ecocentrism" is unfortunate, especially in light of the nuanced distinction drawn between "anthropocentrism," "biocentrism," and "ecocentrism" in Halsey and White (1998).

40. The A.L.F. draws a distinction between its short-term and long-term goals. The A.L.F., whose mission is "[t]o effectively allocate resources (time and money) to end the 'property' status of nonhuman animals" and whose "objective of the mission" is "[t]o abolish institutionalized animal exploitation because it assumes that animals are property," identifies a "short-term aim" of "sav[ing] as many animals as possible and directly disrupt[ing] the practice of animal abuse" and a "long-term aim" of "end[ing] all animal suffering by forcing animal abuse companies out of business" (A.L.F., n.d., *About ALF: Mission Statement*; A.L.F., n.d., *The ALF Credo and Guidelines*).
41. As the Erie "Elves" stated unequivocally in their communiqué, as suggested above (see n.15 and accompanying text), and, as Hays and colleagues contend (1996: 172), "radical environmentalists who participate in sabotage fundamentally distrust the political process." Johnston and Johnston (2020: 576–577) offer a related, but slightly different perspective, concluding that the actions of the RALM are "an expression of [the activists'] deep distrust of the formal justice system, which they understand as failing on several counts—as it is seen neither to protect the innocent and vulnerable nor to punish or curtail the actions of those who would harm and exploit animals, the earth, and their fellow human beings for the sake of profit." (For an examination of how the RALM's perceptions of justice and strategies for resistance intersect with other social issues/social movements, see Johnston and Johnston (2020: 564, 574–575; see also 2017:736); for a note about how "the original Earth First!ers began to engage in a radical form of environmentalism that rejected conventional democratic methods in favor of direct action and even sabotage," see Woodhouse (2018:

5); for a discussion of the E.L.F.'s lack of faith in the legal system, as well as its relationship to mainstream environmental organizations that employ "traditional tactics," see Pickering (2007: 56, 59–61).)

Notwithstanding the statements of Hays and colleagues (1996) and Johnston and Johnston (2020), many activists/ecosaboteurs/eco-warriors do feel that ecodefense has (or may have) strategic value as one of a number or range of tactics (see Chapter 1, n.12 (citing Brown 2021; Burkett 2016; Malm 2021; McIntyre 2021; Smith 2018)) that can help protect the environment effectively. Thus, many such activists/ecosaboteurs/eco-warriors consider their actions as working in conjunction with those of mainstream organizations—in a "bad cop"-"good cop" sense—so that mainstream organizations' demands appear more moderate (see Shevory 1996: 191; Vanderheiden 2005: 437; see generally Woodhouse 2018: 9)—what Andreas Malm, author of *How to Blow Up a Pipeline: Learning to Fight in a World on Fire*, refers to as the "radical flank effect" (2021: 50; see also Klein 2021; Wickenden 2021; Wolfe 2022/2023; for examples of the "radical flank" in environmental campaigns in Australia, Canada, and the United States since the 1970s, see McIntyre 2021: 13, 37, 65, 78, 110, 117, 133, 221, 224; for a discussion of "positive 'radical flank' effects" and "negative 'radical flank' effects," see Farrer and Klein (2022: 218, 219, 221-223); for examples of environmental "victories" achieved through "a diverse tactical mix of [nondestructive/nonviolent] civil disobedience, litigation, and lobbying," see Brown 2021: 360–61 (citing research by Chenoweth and Stephan 2011)).

In response to the question, "Why don't you [the E.L.F.] use legal methods?," Pickering (2007: 26) responds with a question of his own: "How many successful movements in history can you think of off the top of your head that came about only by education? How do you 'protest' against slavery? Holding signs and handing out flyers on southern plantations?"

Similarly, in an interview with Sean Illing for *Vox*, Malm submits that "the past experiences of social struggles suggests [sic] that if you're fighting a very powerful enemy, you need to engage in tactics that can impose costs on that enemy. This usually includes forms of property destruction and confrontation with the ruling order that goes beyond absolutely peaceful civil disobedience" (quoted in Illing 2021). As Malm (quoted in Illing 2021) continues, "I don't know of any relevant analogy or a parallel structure in the past that has succeeded without an element of more militant methods. I don't see how we can imagine that we will win this fight while staying as gentle and kind and polite as we have in the climate movement so far." In *How to Blow Up a Pipeline: Learning to Fight in a World on Fire*, Malm (2021: 23) develops this point further, stating that "[t]hus far, the movement for averting a spiralling climate catastrophe has not only been civil: it has been gentle and mild in the extreme." Malm (2021: 49, 50) contends that "the civil rights movement won the [Civil Rights] Act of 1964 *because it had a radical flank that made it appear as a lesser evil in the eyes of state power*. . . . Next to the threat of black revolution—Black Power, the Black Panther Party, black guerrilla groups—integration seemed a tolerable price to pay. Without Malcolm X, there might not have been a Martin Luther King (and vice versa)" (emphasis in original) (for similar suggestions, see, e.g., Farrer and Klein 2022: 222; Vanderheiden 2005: 435).

Malm's (2021) position seems to echo Churchill's (2017: 68, 69) assertion that "the limited success attained by Martin Luther King Jr. and his disciples in the United States during the 1960s, using a strategy consciously guided by Gandhian principles of nonviolence, owes a considerable debt to the existence of less pacifist circumstances. . . . . [A]ny nonviolent confrontation of state power must ultimately depend either on the state refraining from unleashing some real measure of its potential violence or the active presence of some counterbalancing violence of precisely the sort pacifism professes to reject as a political opinion."

For an argument about the role and significance of international terrorism for (the recognition of) the Palestine Liberation Organization, see Hoffman (2017); see also Bonner (1998); cf. Finlay (2009: 758–760, 763, 767, 773n.90).

For a discussion of the difference between ecosabotage and sabotage for nonecological purposes (e.g., "wartime sabotage"), as well as an explanation of how ecosabotage can be distinguished from vandalism—"the destruction of property or other mischief that is motivated by malice or spite"—see Martin (1990: 293);

cf. Malm 2021: 109). For a brief history of sabotage and case for sabotage as a form of resistance to capitalism and climate change—and means of defending the environment, see Lossin (2018); see also Malm (2021: 68). For an exploration of the impact of "environmental sabotage" on the environmental movement in the mid-1970s to the mid-1990s, see Hayes et al. (1996: 176–178).

References

Abbey, Edward. [1975] 2000. *The Monkey Wrench Gang*. New York: Perennial Classics.

AFP. 2008. Greenpeace Vessel Ends Pursuit of Japanese Whalers. *Terra Daily*, January 26. https://www.terradaily.com/2007/080126094217.ulo9idv9.html (last visited September 2, 2024).

Allhoff, Steffen W., and Sebastian K. Buciak. 2013. Action, Reaction, Escalation: Radical Environmental Organisations in the Context of Eco-terrorism and Eco-radicalism. In *Global Environmental Change: New Drivers for Resistance, Crime and Terrorism?*, Achim Maas, Balázs Bodó, Clementine Burnley, Irina Comardicea, and Roger Roffey, ed. pp. 221–242. Baden-Baden, DE: Nomos Verlagsgesellschaft.

Amster, Randall. 2006. Perspectives on Ecoterrorism: Catalysts, Conflations, and Casualties. *Contemporary Justice Review*, 9(3): 287–301. https://doi.org/10.1080/10282580600827991.

Anderson, Benedict. [1983] 2006. *Imagined Communities: Reflections on the Origin and Spread of Nationalism*. Revised. New York: Verso.

Animal Liberation Front (A.L.F.). n.d. About ALF: Mission Statement. http://www.animalliberationfront.com/ALFront/mission_statement.htm (last visited December 21, 2014).

Animal Liberation Front (A.L.F.). n.d. The ALF Credo and Guidelines. http://www.animalliberationfront.com/ALFront/alf_credo.htm (last visited December 21, 2014).

Associated Press. 2009. Sabotage Fells 2 Radio Towers. *The New York Times*, September 16: 18. https://www.nytimes.com/2009/09/06/us/06towers.html.

Associated Press. 2019. 4 activists arrested in oil pipeline shutdown attempt. *APNews.com*, February 4. Available at: https://apnews.com/general-news-c816b1cf8e1047f688fe349e2712d0f4.

Bandow, Doug. 1991. Radical Activism Cannot Help Protect the Environment. In *The Environmental Crisis: Opposing View Points*, Neal Bernards, ed. pp. 254–260. San Diego, CA: Greenhaven.

Bedau, Hugo A. 1961. On Civil Disobedience. *The Journal of Philosophy*, 58(21) [October 12]: 653–665. https://doi.org/10.2307/2023542.

Best, Steven and Anthony J. Nocella, II, ed. 2004. *Terrorists or Freedom Fighters?: Reflections on the Liberation of Animals*. New York: Lantern Books.

Boggs, Carl. 1977. Marxism, Prefigurative Communism, and the Problem of Workers' Control. *Radical America*, 11(6)/12(1) [Special Double Issue: Winter: November 1977-February 1978]: 99–122. https://repository.library.brown.edu/storage/bdr:89258/pdf/.

Bonner, Raymond. 1998. Getting Attention. *The New York Times*, November 1. https://archive.nytimes.com/www.nytimes.com/books/98/11/01/reviews/981101.01bonnert.html.

Brisman, Avi. 2004. Double Whammy: Collateral Consequences of Conviction and Imprisonment for Sustainable Communities and the Environment. *William & Mary Environmental Law & Policy Review*, 28(2): 423–475.

Brisman, Avi. 2007. Toward a More Elaborate Typology of Environmental Values: Liberalizing Criminal Disenfranchisement Laws and Policies. *New England Journal on Criminal & Civil Confinement*, 33(2): 283–457.

Brisman, Avi. 2008. Crime-Environment Relationships and Environmental Justice. *Seattle Journal for Social Justice*, 6(2): 727–817. https://digitalcommons.law.seattleu.edu/cgi/viewcontent.cgi?article=1138&context=sjsj.

Brisman, Avi. 2009. It Takes Green to Be Green: Environmental Elitism, "Ritual Displays", and Conspicuous Non-Consumption. *North Dakota Law Review*, 85(2): 329–370.

Brisman, Avi, and Alison Rau. 2009. From Fear of Crime to Fear of Nature: The Problem with Permitting Loaded, Concealed Firearms in National Parks. *Golden Gate University Environmental Law Journal*, 2(2): 255–272.

Brooke, James. 1998. Group Claims Responsibility for Blazes At Vail Resort. *The New York Times*, October 22: A14. https://www.nytimes.com/1998/10/22/us/group-claims-responsibility-for-blazes-at-vail-resort.html.

Brown, Charles S. 2007. Beyond Intrinsic Value: Undermining the Justification of Ecoterrorism. *The American Journal of Economics and Sociology*, 66(1): 113–125. http://www.jstor.org/stable/27739623.

Brown, Joseph M. 2020. Notes to the Underground: Credit Claiming and Organizing in the Earth Liberation Front. *Terrorism and Political Violence*, 32(2): 237–256. https://doi.org/10.1080/09546553.2017.1364637. Published online on September 7, 2017.

Brown, Joseph M. 2021. Civil Disobedience, Sabotage, and Violence in US Environmental Activism. In *The Oxford Handbook of Comparative Environmental Politics*, Jeannie Sowers, Stacy D. VanDeveer, and Erika Weinthal, ed. pp. 356–374. Online edition, Oxford Academic. https://doi.org/10.1093/oxfordhb/9780197515037.013.34.

Burkett, Maxine. 2016. Climate Disobedience. *Duke Environmental Law & Policy Forum*, 27(1) [Fall]: 1–50. https://scholarship.law.duke.edu/delpf/vol27/iss1/1/.

Chenoweth, Erica, and Maria J. Stephan. 2011. *Why Civil Resistance Works: The Strategic Logic of Nonviolent Conflict*. New York: Columbia University Press.

Chuk'shown Earth First! 2004. Notorious SAB. *Earth First!*, 24(4) [May-June]: 6–7.

Churchill, Ward. 2017. Pacificism as Pathology: Notes on an American Pseudopraxis. In *Pacifism as Pathology: Reflections on the Role of Armed Struggle in North America*, Ward Churchill and Michael Ryan, 58–139. Oakland, CA: PM Press.

Cianchi, John. 2013. *I Talked to My Tree and My Tree Talked Back: Radical Environmental Activists and Their Relationships With Nature*. Ph.D. Thesis, School of Social Sciences, University of Tasmania (Australia).

Cianchi, John. 2015. *Radical Environmentalism: Nature, Identity and More-than-Human Agency*. Basingstoke, Hampshire, UK: Palgrave Macmillan.

Cohen, Carl. 1971. *Civil Disobedience: Conscience, Tactics, and the Law*. New York: Columbia University Press.

Combs, Cindy C. 2006. *Terrorism in the Twenty-First Century*, 4th edn. Upper Saddle River, NJ: Pearson Prentice Hall.

Cottee, Simon, and Keith Hayward. 2011. Terrorist (E)motives: The Existential Attractions of Terrorism. *Studies in Conflict and Terrorism*, 34(12): 963–986. https://doi.org/10.1080/1057610X.2011.621116.

Cox, Jackie, and Pip Donovan. 2020. *Counting the Crimes*. A report by Action Against Foxhunting, August. https://www.actionagainstfoxhunting.org/wp-content/uploads/2020/08/Counting-the-Crimes-Final-310720.pdf.

Curnow-Garland, Davey. 2010. Europe: 40 Years of Ecological Resistance & Protest. *Resistance, Journal of the Earth Liberation Movement*, Spring: 28–31.

Davis, John, ed. 1995. *The Earth First! Reader: Ten Years of Radical Environmentalism*. Bombay, India: International Institute for Sustainable Future.

Dereniowska, Małgorzata, and Jason P. Matzke, Jason P. 2018. On Compromise in Radical Environmental Activism [O kompromisie w radykalnym aktywizmie środowiskowym]. *Humanistyka I Przyrodoznawstwo*, 24: 9–38. https://doi.org/10.31648/hip.2595; https://czasopisma.uwm.edu.pl/index.php/hip/article/view/2595/2134; https://cejsh.icm.edu.pl/cejsh/element/bwmeta1.element.desklight-355ea4ac-ee7a-4c1e-9a8d-52c61029c4e2/c/2018_1_Dereniowska.pdf.

Devall, Bill, and George Sessions. 1985. *Deep Ecology: Living as if Nature Mattered*. Salt Lake City, UT: Gibbs M. Smith Inc.

Dickel, Sascha. 2012. Beyond Cynicism: Engineering Technologies and Vegan Diets as Alternative Solutions for Climate Change. In *Global Environmental Change: New Drivers for Resistance, Crime and Terrorism?*, Achim Maas,

Balázs Bodó, Clementine Burnley, Irina Comardicea, and Roger Roffey, ed. pp. 243–259. Baden-Baden, DE: Nomos Verlagsgesellschaft.

Drengson, Alan R., and Yuichi Inoue, ed. 1995. *The Deep Ecology Movement: An Introductory Anthology*. Berkeley, CA: North Atlantic Books.

Dryzek, John. 2000. *Deliberative Democracy and Beyond: Liberals, Critics, Contestations*. Oxford: Oxford University Press.

Earth First! n.d. *About Earth First!* http://www.earthfirst.org/about.htm (last visited December 21, 2014).

Farrer, Benjamin, and Graig R. Klein. 2017. The Political Roots of Domestic Environmental Terrorism. *Journal of Elections, Public Opinions, and Parties*, 27(2): 133–155. https://doi.org/10.1080/17457289.2016.1247846.

Farrer, Ben, and Graig R. Klein. 2022. How Radical Environmental Sabotage Impacts US Elections. *Terrorism and Political Violence*, 34(2): 218–239. https://doi.org/10.1080/09546553.2019.1678468.

Federman, Adam. 2020. Revealed: US listed climate activist group as 'extremists' alongside mass killers. *The Guardian*, January 13. https://www.theguardian.com/environment/2020/jan/13/us-listed-climate-activist-group-extremists.

Finlay, Christopher J. 2009. How to do Things with the Word 'Terrorist.' *Review of International Studies*, 35(4) [October]: 751–774. http://www.jstor.org/stable/40588072.

Flükiger, Jean-Marc. 2009. The Radical Animal Liberation Movement: Some Reflections on Its Future. *Journal for the Study of Radicalism*, 2(2): 111–132. https://doi.org/10.1353/jsr.0.0000. Available at: https://muse.jhu.edu/article/254911.

Flynn, Jim. 2010. The Grassroots Contingent of the Environmental Movement. *Resistance, Journal of the Earth Liberation Movement*, Spring: 21–23.

Foreman, Dave. 1991. *Confessions of an Eco-Warrior*. New York: Crown.

Foreman, Dave, and Bill Haywood, ed. 1993. *Ecodefense: A Field Guide to Monkeywrenching*, 3rd edn. Chico, CA: Abbzug Press.

Fountain, Henry. 2006. Rising Above the Environmental Debate. *The New York Times*, June 18. https://www.nytimes.com/2006/06/18/weekinreview/18basic.html?_r=0.

Glick, Daniel. 2003. *Powder Burn: Arson, Money, and Mystery on Vail Mountain*. New York: PublicAffairs.

Gray-Donald, David. 2015a. Canadian Activists Show There's a Very Easy (But Dangerous) Way to Shut Down Pipelines. *Vice*, December 17. https://www.vice.com/en/article/canadian-activists-show-theres-a-very-easy-but-dangerous-way-to-shut-down-pipelines/.

Gray-Donald, David. 2015b. Canadian Protesters Keep Shutting Down the Line 9 Oil Pipeline. *Vice*, December 22. https://www.vice.com/en/article/protesters-keep-shutting-down-the-line-9-oil-pipeline/.

Haiven, Max, and Alex Khasnabish. 2014. *The Radical Imagination: Social Movement Research in the Age of Austerity*. London: Zed Books.

Hale, Mike. 2010. In the High Seas, a Mix of Danger and Bumbling High Jinks. *The New York Times* (New York edition), August 27: C4. Published online as "On the High Seas, a Mix of Danger and Bumbling High Jinks" on August 26, 2010, at https://www.nytimes.com/2010/08/27/arts/television/27whale.html.

Hasler, Olivia, Reece Walters, and Rob White. 2020. In and Against the State: The Dynamics of Environmental Activism. *Critical Criminology: An International Journal*, 28(3): 517–531. https://doi.org/10.1007/s10612-019-09432-0.

Halsey, Mark, and Rob White. 1998. Crime, Ecophilosphy and Environmental Harm. *Theoretical Criminology*, 2(3) [August]: 345–371. https://doi.org/10.1177/1362480698002003003.

Hays, Scott, Michael Esler, and Carol Hays. 1996. Radical Environmentalism and Crime. In *Environmental Crime and Criminality: Theoretical and Practical Issues*, Sally M. Edwards, Terry D. Edwards, and Charles B. Fields, ed. pp. 163–181. New York and London: Garland Publishing.

Hill, Julia Butterfly. 2000. *The Legacy of Luna: The Story of a Tree, a Woman and the Struggle to Save the Redwoods*. San Francisco, CA: Harper San Francisco.

Hoffman, Bruce. 2017. *Inside Terrorism*, 3rd edn. New York: Columbia University Press.

Holthaus, Eric. 2019. Valve Turners Try to Shut Off Minnesota Pipelines, Say 'Politicians Won't Act.' *Grist*, February 5. https://grist.org/article/valve-turners-try-to-shut-off-minnesota-pipelines-say-politicians-wont-act/.

Illing, Sean. 2021. The Case for a More Radical Climate Movement: Author Andreas Malm on the failures of climate activism and the need for escalation. *Vox*, October 1. https://www.vox.com/vox-conversations-podcast/22691428/vox-conversations-climate-change-andreas-malm.

Johnson, Erik W., Jon Agnone, and John D. McCarthy. 2010. Movement Organizations, Synergistic Tactics and Environmental Public Policy. *Social Forces*, 88(5) [July]: 2267–2292. https://doi.org/10.1353/sof.2010.0038

Johnston, Genevieve, and Matthew S. Johnston. 2017. 'We Fight for all Living Things': Countering Misconceptions About the Radical Animal Liberation Movement. *Social Movement Studies*, 16(6): 735–751. https://doi.org/10.1080/14742837.2017.1319268.

Johnston, Genevieve, and Matthew S. Johnston. 2020. 'Until Every Cage is Empty': Frames of Justice in the Radical Animal Liberation Movement. *Contemporary Justice Review*, 23(4): 563–580. https://doi.org/10.1080/10282580.2020.1719361.

Jones, Pattrice. 2004. Mothers with Monkeywrenches: Feminist Imperatives and the ALF. In *Terrorists or Freedom Fighters?: Reflections on the Liberation of*

Animals, Steven Best and Anthony J. Nocella, II, ed. pp. 137–156. New York: Lantern Books.

Joosse, Paul. 2007. Leaderless resistance and ideological inclusion: The case of the Earth Liberation Front. *Terrorism and Political Violence*, 19(3): 351–368. https://doi.org/10.1080/09546550701424042.

Joosse, Paul. 2012. Elves, Environmentalism, and "Eco-terror": Leaderless Resistance and Media Coverage of the Earth Liberation Front. *Crime Media Culture*, 8(1): 75–93. https://doi.org/10.1177/1741659011433366.

Joosse, Paul. 2014. Antiglobalization and Radical Environmentalism: An Exchange on Ethical Grounds. *Ethics in Progress*, 5(1): 33–51. https://doi.org/10.14746/eip.2014.1.2; https://pressto.amu.edu.pl/index.php/eip/article/view/9382/9084.

Joosse, Paul. 2021. Narratives of rebellion. *European Journal of Criminology*, 18(5): 735–754. https://doi.org/10.1177/1477370819874421.

Katz, Jack. 1988. *Seductions of Crime: Moral and Sensual Attractions in Doing Evil*. New York: Basic Books.

Klein, Ezra. 2021. It Seems Odd That We Would Just Let the World Burn. *The New York Times*, July 15. https://www.nytimes.com/2021/07/15/opinion/climate-change-energy-infrastructure.html.

Kuipers, Dean. 2018. Pipeline Vandals Are Reinventing Climate Activism. *WIRED*, November 9. https://www.wired.com/story/monkeywrenching-vandals-are-reinventing-climate-activism/.

Lemons, John, and Donald A. Brown. 2011. Global climate change and non-violent civil disobedience. *Ethics in Science and Environmental Politics*, 11(1): 3–12. https://doi.org/10.3354/esep00109; https://www.int-res.com/articles/esep2011/11/e011p003.pdf.

Liddick, Donald R. 2006. *Eco-Terrorism: Radical Environmental and Animal Liberation*. Westport, CT: Praeger.

Liddick, Don. 2013. Techniques of Neutralization and Animal Rights Activists. *Deviant Behavior*, 34(8): 618–634. https://doi.org/10.1080/01639625.2012.759048.

Loadenthal, Michael. 2017. "Eco-Terrorism": An Incident-Driven History of Attack (1973–2010). *Journal for the Study of Radicalism*, 11(2) [Fall]: 1–34. https://www.jstor.org/stable/26297171.

Lossin, R.H. 2018. Sabotage as Environmental Activism. *Public Seminar*, July 3. https://publicseminar.org/essays/sabotage-as-environmental-activism/.

Malm, Andreas. 2021. *How to Blow Up a Pipeline: Learning to Fight in a World on Fire*. London and New York: Verso.

Manson, Dylan. 2024. Eco-sabotage as Defensive Activism. *Ethical Theory and Moral Practice*, 27(4) [September]: 505–522. https://doi.org/10.1007/s10677-024-10449-w.

Martin, Michael. 1990. Ecosabotage and Civil Disobedience. *Environmental Ethics*, 12(4): 291–310.

Martin, Glen. 1998. A YEAR IN THE SKY/Last December 10, Julia Butterfly climbed 180 feet up an ancient redwood tree in Humboldt County's beleaguered Headwaters Forest, named the tree Luna, and refused to come down. She's still up there. *San Francisco Chronicle/SFGate*, December 6. Available at: https://www.sfgate.com/news/article/a-year-in-the-sky-last-december-10-julia-2974427.php.

McIntyre, Iain. 2021. *Environmental Blockades: Obstructive Direct Action and the History of the Environmental Movement*. Abingdon, Oxon, UK, and New York: Routledge.

Mills, Stephanie. 1991. Thoughts from the RRR. In *The Earth First! Reader: Ten Years of Radical Environmentalism*, John Davis, ed. pp. 128–132. Salt Lake City, UT: Peregrine Smith.

Moran, Grey. 2018. Here's what climate scientist James Hansen would have said in the Valve Turner trial. *Grist*, October 18. https://grist.org/article/heres-what-climate-scientist-james-hansen-would-have-said-in-the-valve-turner-trial/.

Moss, Michael, and Souad Mekhennet. 2007. A look at the playbook for Islamic militants. *The New York Times*, June 10. https://www.nytimes.com/2007/06/10/world/africa/10iht-jihad.2.6075853.html.

Næss, Arne. 1973. The Shallow and the Deep, Long-Range Ecology Movement. A Summary. *Inquiry*, 16(1–4): 95–100. Published online on August 29, 2008. https://doi.org/10.1080/00201747308601682; http://www.fraw.org.uk/data/ap/naess_1972.pdf.

Næss, Arne. 1984. A Defence of the Deep Ecology Movement. *Environmental Ethics*, 6(3) [Fall]: 265–270. https://doi.org/10.5840/enviroethics19846330.

Næss, Arne. 1986. The Deep Ecological Movement: Some Philosophical Aspects. *Philosophical Inquiry*, 8(1/2) [Winter]: 10–31. https://doi.org/10.5840/philinquiry198681/22.

Nicholson, Blake. 2019. Companies Decry 'Valve Turners' who Shut Down Pipelines. *APNews.com*, March 9. https://apnews.com/article/0fe28195510141eea0e407687fd49ce5.

Nijhuis, Michelle. 2018. The Valve Turners. *The New York Times Magazine*, February 18: 42. Published online as "'I'm Just More Afraid of Climate Change Than I Am of Prison'" on February 13, 2018, at https://www.nytimes.com/2018/02/13/magazine/afraid-climate-change-prison-valve-turners-global-warming.html.

North American Animal Liberation Press Office (NAALFPO). 2008. Multiple Targets Locks Glued Tires Slashed. *North American Animal Liberation Press Office*, December 8. https://animalliberationpressoffice.org/NAALPO/2008/12/08/multiple-targets-locks-glued-tires-slashed/ (last visited September 2, 2024; on file with author). Originally published as "Communique from the Animal Liberation Front" on December 8, 2008, at http://www.animalliberationpressoffice.org/communiques/2008/2008-12-08_multiple.htm (last visited September 15, 2009; on file with author).

North American Animal Liberation Press Office (NAALFPO). 2009. 5 Telephones from the Biocidal TELMEX Company Sabotaged in Guadalajara, Mexico. *North American Animal Liberation Press Office*, July 10. https://animalliberationpressoffice.org/NAALPO/2009/07/10/5-telephones-from-the-biocidal-telmex-company-sabotaged-in-guadalajara-mexico/ (last visited September 2, 2024; on file with author). Originally published as "Communique from the Environmentalist Combatants for the Liberation of the Earth" on July 10, 2009, at http://www.animalliberationpressoffice.org/communiques/2009/2009-07-10_telmex.htm (last visited September 15, 2009; on file with author).

North American Earth Liberation Front Press Office (NAELFPO). 2000. Communique · 03.24.00. March 24. Accessed at: http://www.elfpressoffice.org/comm032400.html (last visited September 15, 2009; on file with author).

North American Earth Liberation Front Press Office (NAELFPO). 2003. Communique · 01.01.03. January 1. http://www.elfpressoffice.org/comm010103.html (last visited September 15, 2009; on file with author).

North American Earth Liberation Front Press Office (NAELFPO). 2008. Earth Liberation Front Press Office Media Release Archive: North American Earth Liberation Front Press Office Returns. October 31. http://www.elfpressoffice.org/naelfpo103108.pdf (last visited September 15, 2009; on file with author).

North American Earth Liberation Front Press Office (NAELFPO). 2009. Media Release: Earth Liberation Front Topples Radio Station Towers in Snohomish County, WA. September 4. http://www.elfpressoffice.org/release090409html (last visited September 15, 2009; on file with author).

North American Earth Liberation Front Press Office (NAELFPO). n.d. Earth Liberation Front Diary of Actions · 2005. http://www.elfpressoffice.org/actions2005.html (last visited September 15, 2009; on file with author).

North American Earth Liberation Front Press Office (NAELFPO). n.d. Frequently Asked Questions About the Earth Liberation Front. http://www.elfpressoffice.org/elffaqs.html (last visited September 10, 2009; on file with author; many of these “FAQs” are reproduced in Pickering (2007: 25–26, 42–64)).

Oak, Chessie. 2022. Tree Sit Blockades and Queer Liberation. In *Y'all Means All: The Emerging Voices Queering Appalachia*, Z. Zane McNeil, ed. pp. 110–116. Oakland, CA: PM Press.

Oldenburg, Don. 2004. Julia Butterfly Hill, From Treetop to Grass Roots. *The Washington Post*, September 22. https://www.washingtonpost.com/archive/lifestyle/2004/09/22/julia-butterfly-hill-from-treetop-to-grass-roots/128207f1-a301-4e8c-8d87-bdd72f9ec1bb/; https://www.washingtonpost.com/wp-dyn/articles/A40230-2004Sep21_2.html.

Pellow, David Naguib. 2014. *Total Liberation: The Power and Promise of Animal Rights and the Radical Earth Movement*. Minneapolis, MN: University of Minnesota Press.

Pickering, Leslie James. 2007. *The Earth Liberation Front, 1997–2002*, 2nd edn. Portland, OR: Arissa Media Group.

Puckett, Karl. 2020. Montana Supreme Court rejects Oregon Man's Defense That Climate Change Threat Justified Actions. *Great Fall Tribune*, March 4. https://www.greatfallstribune.com/story/news/2020/03/04/climate-change-defense-rejected-montana-supreme-court-leonard-higgins/4953772002/.

Rawls, John. 1999. *A Theory of Justice*. Revised. Cambridge, MA: The Belknap Press of Harvard University Press.

Resistance Magazine. 2009a. Reflections on the Green Scare: What we've learned. *Resistance, Journal of the Earth Liberation Movement*, Summer: 8–11 (on file with author).

Resistance Magazine. 2009b. Earth Liberation: Movement Strategy & Direction. *Resistance, Journal of the Earth Liberation Movement*, Summer: 14–15, 17 (on file with author).

Resistance Magazine. 2009c. Editor's Note. *Resistance, Journal of the Earth Liberation Movement*, Fall: 5 (on file with author).

Resistance Magazine. 2009d. On the Frontlines: With Captain Paul Watson and the Sea Shepherd Conservation Society. *Resistance, Journal of the Earth Liberation Movement*, Fall: 14–19, 31, 35 (on file with author).

Rosebraugh, Craig. 2004. *Burning Rage of a Dying Planet: Speaking for the Earth Liberation Front*. New York: Lantern.

Salter, Colin. 2011. Activism as Terrorism: The Green Scare, Radical Environmentalism and Governmentality. *Anarchist Developments in Cultural Studies*, No. 1(2011): Ten Years After 9/11: An Anarchist Evaluation: 211–238. https://journals.uvic.ca/index.php/adcs/article/view/17126/7340.

Scarce, Rik. 2006. *Eco-Warriors: Understanding the Radical Environmental Movement*, Updated edn. Walnut Creek, CA: Left Coast Press.

Sea Shepard Conservation Society. n.d. Crewing at Sea FAQs. https://seashepherd.org/crewing-at-sea-faq/ (last visited September 4, 2024).

Sea Shepherd Conservation Society. n.d. Who We Are: Sea Shepherd's Mission Statement. http://www.seashepherd.org/who-we-are/ (last visited December 20, 2014); also available at: https://aileengettyfoundation.org/sea-shepherd/ (last visited August 31, 2024).

Shevory, Thomas C. 1996. Monkeywrenching: Practice in Search of a Theory. In *Environmental Crime and Criminality: Theoretical and Practical Issues*, Sally M. Edwards, Terry D. Edwards, and Charles B. Fields, ed. pp. 183–204. New York and London: Garland Publishing.

Smallwood, Scott. 2005. Speaking for the Animals, or the Terrorists? *The Chronicle of Higher Education*, 51(48) [August 5]: A.8.

Smith, William. 2018. Disruptive Democracy: The Ethics of Direct Action. *Raisons Politiques*, 2018/1(69): 13–27. https://shs.cairn.info/journal-raisons-politiques-2018-1-page-13?lang=en.

Sparrow, Rob. 1997. Anarchist Politics & Direct Action. *The Anarchist Library*. https://theanarchistlibrary.org/library/rob-sparrow-anarchist-politics-direct-action; http://www.spunk.org/texts/intro/sp001641.html.

Stephens-Griffin, Nathan. 2023. Biting Back: A Green-Cultural Criminology of Animal Liberation Struggle as Constructed Through Online Communiques. *Crime Media Culture*, 19(2): 252–273. https://doi.org/10.1177/17416590221110118.

Stephenson, Wen. 2018. This Landmark Trial of Climate Activists Puts the Political System Itself on Trial. *The Nation*, October 5. https://www.thenation.com/article/archive/this-landmark-trial-of-climate-activists-puts-the-political-system-itself-on-trial/.

Taylor, Bron. 2013. Resistance: Do the Ends Justify the Means? *Earth Island Journal*, April 26. https://www.earthisland.org/journal/index.php/articles/entry/resistance_do_the_ends_justify_the_means/.

Vanderheiden, Steve. 2005. Eco-terrorism or Justified Resistance? Radical Environmentalism and the "War on Terror." *Politics & Society*, 33(3): 425–447. https://doi.org/10.1177/0032329205278462.

Wickenden, Dorothy. 2021. Andreas Malm on the Environmental Movement and "Intelligent Sabotage." *The New Yorker*, September 27. https://www.newyorker.com/podcast/political-scene/andreas-malm-on-the-environmental-movement-and-intelligent-sabotage.

Woodhouse, Keith Makato. 2018. *The Ecocentrists: A History of Radical Environmentalism*. New York: Columbia University Press.

Wolfe, Matthew. 2022/2023. Backfire. *The New York Times Magazine*, May 29: 38. Published online as "The Rise and Fall of America's Environmentalist Underground" on May 26, 2022 [updated June 15, 2023], at https://www.nytimes.com/2022/05/26/magazine/earth-liberation-front-joseph-mahmoud-dibee.html.

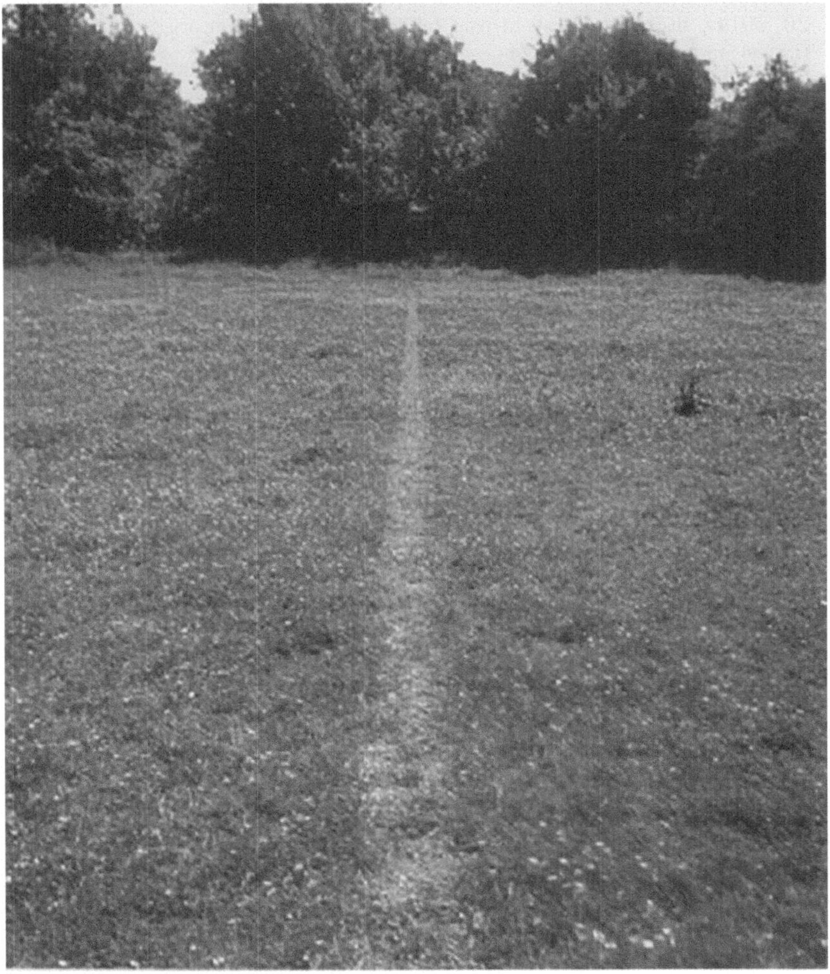

Fig. 1 A Line Made by Walking, 1967, Richard Long. © Richard Long. DACS/Copyright Agency, 2024

CHAPTER 3

Conceptual Art: The Manifestation of an Idea and the Documentation of Art, an Event, or (a) Process

In the previous chapter, this book attempted to better understand ecodefense by exploring the relationship of the communiqué to the action and by comparing a small sample of specific actions and communiqués to each other. In this chapter, I consider the use of documentary information of the Conceptual art of the late 1960s and early 1970s—written declarations, which, I suggest, resemble the style, form, and arguably, in some instances, the purpose of communiqués for ecodefense. I begin with some general comments about Conceptual art and then turn to a brief description of the work and orientations of five artists: Sol LeWitt (1928–2007), Lawrence Weiner (1942–2021), Chris Burden (1946–2015), Hamish Fulton (1946-) and Richard Long (1945-). From here, I offer a couple of observations about the written features of Conceptual art. I conclude this part by describing how a consideration of Conceptual artists' use of written documentation and statements can shed light on the meaning of ecodefense, thereby completing this book's evaluation of the analogy:

> **act of ecodefense: communiqué:: work of conceptual art: written declaration/statement**

The phrase, "conceptual art," is often employed to refer to art in which the concept(s) or idea(s) of or involved in the work are prioritized or

A. Brisman, *Direct Action as Conceptual Art?*, Palgrave Studies in Crime, Media and Culture,
https://doi.org/10.1007/978-3-031-76634-3_3

take precedence over the more traditional aesthetic and material concerns associated with sculpture and painting.[1] While the word, "conceptualist," tends to be applied to "an artist who trades as much in ideas as in materials." (Cotter 2009: C1; see also Rosenberg 2009: C5), the "phenomenon" referred to as "Conceptual art" (or "Conceptual Art," with both words capitalized)—critics (have) tend(ed) to eschew referring to it as a "movement" (see, e.g., Lippard [1973] 1997; Morgan 1994: xiii; 27, 115)—"reached both its apogee and its crisis in the years 1966–72" (Godfrey 1998: 6). Owing much to the French artist, Marcel Duchamp (1887–1968), of the early twentieth century, who famously placed a urinal on its back on a plinth, signed it "R. Mutt," and offered it as a work of art (entitled *Fountain*) to the 1917 exhibition of the Society of Independent artists in New York (Bird and Newman 1999: 2; Godfrey 1998: 6; Meyer 1972: IX-X; Morgan 1994: 5, 1996:9; see also Dutton 2009), the Conceptual artists of the 1960s and 1970s and their progeny privileged and continue to promote ideas; materials—if employed at all—were or are considered or treated as "secondary, lightweight, ephemeral, cheap, unpretentious" (Lippard [1973] 1997: vii; see also Godfrey 1998: 14).[2] Indeed, Kirby (1994: ix) suggests, one might define "Conceptual Art" as "art that strives toward the absence of physicality." Or, as Ursula Meyer (1972: VIII–IX, XIII, XV), the German-born American artist and professor of sculpture, explains:

> Conceptual Art makes the ideational premise of the work known, a decided contrast to other contemporary art, which is not concerned with defining the intention of the work, attending (almost) exclusively to its *appearance*. The *IDEA* of the work, which on the artist could reveal, remains hidden, thus becoming everybody's guessing game and/or responsibility.
>
> . . .
>
> Conceptual Art emphasizes the elimination of the art-object.
>
> . . .
>
> The abolition of the art-object, typical for Conceptual Art, eliminates the concern with 'style,' 'quality,' and 'permanence,' the indispensable modalities of traditional and contemporary art. Style, quality, and permanence support the notion that age is a basic consideration for the value of an art-object.

While the term, "Conceptual," as applied to the visual arts, has been attributed to Ed Keinholz (in the late 1950s) and to Henry Flynt (in 1961) (Bird and Newman 1999: 3; see also Hopkins 2000: 177; Morgan

1994: 13, 118; 1996: 31), the term did not really enter art-world discourse until Sol LeWitt's "Paragraphs on Conceptual Art," published in *Artforum* in 1967; (for a discussion, see, e.g., Fineberg 1995: 306; Godfrey 1998: 12–13; Osborne 1999: 52–56). As LeWitt (1967) set forth:

> I will refer to the kind of art in which I am involved as conceptual art. In conceptual art the idea or concept is the most important aspect of the work. When an artist uses a conceptual form of art, it means that all of the planning and decisions are made beforehand and the execution is a perfunctory affair. The idea becomes a machine that makes the art. This kind of art is not theoretical or illustrative of theories; it is intuitive, it is involved with all types of mental processes and it is purposeless. It is usually free from the dependence on the skill of the artist as a craftsman. It is the objective of the artist who is concerned with conceptual art to make his work mentally interesting to the spectator, and therefore usually he would want it to become emotionally dry. There is no reason to suppose, however, that the conceptual artist is out to bore the viewer. It is only the expectation of an emotional kick, to which one conditioned to expressionist art is accustomed, that would deter the viewer from perceiving this art.
>
> . . .
>
> What the work looks like isn't too important.
>
> . . .
>
> The work does not necessarily have to be rejected if it does not look well.
>
> . . .
>
> If the artist carries through his idea and makes it into visible form, then all the steps in the process are of importance. The idea itself, even if not made visual, is as much a work of art as any finished product.

For LeWitt, who continued to subscribe to these beliefs and make work according to these parameters until his death in 2007, "the concept behind the work actually constitutes the art" (Godfrey 1998: 142) or, as Morgan (1994: 35) puts it, "the manifestation of the idea—his idea or system of instructional notation—[became] the work; the documentation simply record[ed] the process." LeWitt was so committed to the ascendancy of *the idea* that he often did not execute his works himself. Thus, for example, a draughtsman might be given the following instructions: "'Ten thousand random straight lines drawn by one draughtsman, 1,000 lines a day, for ten days, within a 120″ square.' 'Lines, not short, not straight, crossing and touching, drawn at random, using four colours (yellow, black, red and blue), uniformly dispersed with maximum density covering the entire surface of the wall'" (Godfrey 1998: 152). While LeWitt maintained ([1969] 1972: 174) that "[i]deas alone can be works of art; they are in a chain of development that may eventually find some form. All ideas need not be made physical," many of his "wall drawings," which were executed by others pursuant to his predetermined directions, were often quite beautiful and visually compelling (Godfrey 1998: 389; see also Rorimer 1999: 11; Vogel 2009). As Morgan (1994: xiv) explains, LeWitt "accept[ed] the language paradigm as an equivalent to the art without relinquishing the object."

Lawrence Weiner, a contemporary of LeWitt, *was* willing to "relinquish the object." Weiner claimed that the work "'can be presented just in language'" (quoted in Archer 2002: 785)—"that language, being an independent representational system, could exist on its own as the material of a work's construction and therefore that it was possible to forgo the physical construction of any particular work" (Rorimer 1999: 12). Indeed, in a conversation in 1969 with Meyer, Weiner announced, "I do not mind objects, but I do not care to make them" (1972: 217). Or, as Kennedy (2007: AR31) describes in a review, more than forty years ago, "Weiner decided that words would serve almost exclusively as raw material for his art: words spoken, sung, painted on walls, printed in books and on matchbooks, stamped on coins or manhole covers or elsewhere." Indeed, in 1968, Weiner wrote in a sort of "declaration of principles" or "statement of intent" that:

> 1. The artist may construct the piece.
> 2. The piece may be fabricated.
> 3. The piece need not be built.
>
> Each being equal and consistent with the intent of the
> artist, the decision as to condition rests with the receiver upon the occasion of receivership.[3]

As Rorimer (1999: 13) explains,

> [b]y means of this statement, the artist stipulates that the actual, physical realization of any one of his works is not a requirement but an option left open to the discretion of any perceiving subject, including the artist. *RESIDUE OF A FLARE ON A BOUNDARY* (1969), for example, was implemented by Weiner under the auspices of the Stedelijk Museum,

> Amsterdam, during the exhibition *Op Losse Schroeven:Situaties en Cryptostructuren (Square Pegs in Round Holes: Structures and Cryptostructures)* in the spring of 1969. The same year, *A SQUARE REMOVAL FROM A RUG IN USE* (1969) was physically constructed in the home of collectors who had purchased the piece. At their request, Weiner came to their residence in Cologne, Germany, and cut a square section of the rug. He thereby created a lacuna rather than, more typically, adding a new *objet d'art* to their collection.

Similarly, Kennedy (2007: AR31) reveals, how

> when Mr. Weiner made early artworks that consisted of short descriptive phrases like "Two Minutes of Spray Paint Directly Upon the Floor From a Standard Aerosol Spray Can" or "A 36" x 36" Removal to the Lathing or Support Wall of Plaster or Wallboard from a Wall," the work could remain solely as words written on walls or pieces of paper; or Mr. Weiner could create the situations the words described; or someone who bought the pieces, or a museum featuring them in a show, could enact the words with a spray can or a jigsaw and some elbow grease. Or not.

To demonstrate the extent to which Weiner privileged language—and his determination to "'spend the rest of his existence dealing with the 'general idea of materials rather than the specific'" (quoted in Godfrey 1998: 165)—consider his 1968 "exhibition," which consisted of twenty-eight phrases, one per page, as a book entitled *Statements* (for a discussion, see Godfrey 1998: 166; see generally Meyer 1972: XIV).[4] One of the statements appeared as follows:

> A series of stakes set in the ground
> at regular intervals to form a rect
> angle
> Twine strung from stake to stake to
> demark a grid[5]

This statement had been "realized" the previous year (1967) on the grounds of (the now defunct) Windham College, in Putney, Vermont, as part of a sculpture exhibition (see Alberro and Zimmerman 1998: 44–46). As Godfrey (1998: 166) recounts, "[t]he twine had got in the way of the fraternity boys who wanted to play touch football, so they cut it.

Weiner realized that it ultimately did not matter: the words or instructions were sufficient."

The relationship of (*the* or *a work of*) art to written declarations, descriptions, instructions, or statements differed for Chris Burden, a California-based artist, who in 1976, exhibited a number of "relics" at the Ronald Feldman Gallery in New York. *Trans-fixed*, for example, consisted of two blood-stained nails accompanied by the following text:

> TRANS-FIXED, Venice, California, April 23, 1974. Inside a small garage on Speedway Avenue, I stood on the rear bumper of a Volkswagen. I lay on my back over the rear section of the car, stretching my arms onto the roof. Nails were driven through my palms into the roof of my car. The garage door was opened and the car was pushed halfway out into Speedway. Screaming for me, the engine was run at full speed for two minutes. After two minutes the engine was turned off, and the car pushed back into the garage. The door was closed.[6]
>
> (Fineberg 1995: 342; see also Kaye 2018: 28)

For Burden, the description and nails were not (the) "art." The "art" occurred on April 23, 1974. The nails and description were "artifacts of this machine-age crucifixion"—nothing more (Fineberg 1995: 342).[7]

Other work by Burden from this period resembled *Trans-fixed*. In *Shoot*, performed at F-Space gallery, Santa Ana, California, in 1971, Burden arranged for a friend to shoot him (Hopkins 2000: 191; Plagens 1973; see also Arn 2023; Ghorashi 2015; Kennedy 2013; Kutner 2015).[8] The piece did not unfold exactly as planned (see Arn 2023; *BBC News* 2017; Boucher 2017; Fox 2015). As Burden explains, "[y]ou saw a lot of people getting shot on TV every night in Vietnam. Guys my age" (quoted in Ghorashi 2015). The best possible outcome for the piece, according to Burden, was that "[t]he bullet would whiz by my arm, and it would scratch it, and one drop of blood would roll down my forearm. That was the ideal-ideal" (quoted in Ghorashi 2015). Instead—"either the shooter or the artist tensed up slightly" (*BBC News* 2017)—Burden sustained a much deeper wound in his arm, which required medical assistance (Arn 2023; *BBC News* 2017; Fox 2015; Ghorashi 2015). While Burden had expected the bullet to graze his arm—he later described how he "felt like a truck [had] hit [his] arm at 80 miles per hour" (quoted in Fox 2015)—he did not regard the fact that the piece did not transpire as intended to be particularly consequential. As Boucher (2017) explains, "[d]espite the mishap, Burden thought the work held up a mirror to America. 'It's as American as apple pie,' he [Burden] stated, 'either being shot or shooting people.'" Burden subsequently exhibited a blurry, black-and-white photograph, in which one can barely discern two individuals, one with a rifle. The following label accompanied the image:

> At 7:45 p.m. I was shot in the left arm by a friend. The bullet was a copper jacket 22 long rifle. My friend was standing about fifteen feet from me.
>
> (Burden 1974: 24; see also O'Dell 1998: 1)

With *Shoot*, the photograph and text did not constitute the art. The art took place at 7:45 pm on November 19, 1971, in F Space; the photograph and text were merely evidence of the art.[9]

Not all Conceptual art involving the body entailed such startling acts of self-mutilation. For instance, British artists, Hamish Fulton and Richard Long, both made—and have continued to make—art by going for walks[10]—or what Lorimer (2016: 26) terms "footworks," what Robinson (2013: 16) labels "walk-works," at least with respect to Fulton, and what Macfarlane (2009) calls "art *walks* rather than art *works*" (emphasis added), when describing Long's perambulations.[11] Fulton has taken walks and then exhibited in art galleries large, grainy photographs with a line or two of text describing something about the conditions of the walk or something about his experience while walking (Morgan 1996: 83)—"[t]he fullness of [his] experience... stripped-down to a graphic or typographical summation for exhibition and interpretation" (Lorimer 2016: 26).

Similarly, Long has gone for a walk and then exhibited "a map with the route of the walk drawn on it, a text listing things passed or even seen *en route*, a photograph, a tabulation of objects spotted and/or carried on the walk, and so on" (Archer 2002: 88; see also Bird and Newman 1999: 4; Godfrey 1998: 203, 430; Wood 1999: 66).[12] Occasionally, Long has stopped to arrange stones or twigs into a pattern and then photograph the configuration (Archer 2002: 88; see also Godfrey 1998: 226; Higgins 2012; Jones 2009b; Macfarlane 2009; see generally Brown 2009; Fuchs 1986: 72; Jones 2009a; O'Hagan 2009; Seymour 1991). Wallis (2009: 33) characterizes this work as "interventions in the landscape," and Collier and Morrison-Bell (2013: 88) refer to this "rearranging [of] natural elements" as "marks on the world." Fuchs (1986: 72, 43) uses similar phrasing to that of Collier and Morrison-Bell, describing Long's art as "leaving... marks on the surface of the earth," but then, elsewhere, seems to prefer to think of Long's works as "*traces* of staying and passing" (emphasis added). Meanwhile, Lorimer (2016: 26) makes clear that these really are "*respectful* alteration[s] of the landscape" (emphasis added)—a comment that reflects Jones' (2009b) estimation that Long leaves only "ephemeral marks... gentle marks." Per Jones (2009b), Long "set out, in the 1960s, to relate to nature in a new way, making art in the landscape without scarring it. A deep respect is manifest in everything he [Long] does." Fuchs (1986: 45) stresses that "[i]mpermanence lies at the very heart of... Long's conviction as an artist"—and, indeed, Long (2002 [1997]: 146) is adamant that it has "never my intention to make permanent or monumental works in the landscape"[13]—an ethos that would resonate with ecodefenders.

Because members of the public do/did not accompany Fulton or Long on their walks, they cannot/could not actually experience the(ir) art; they can/could view only ***the documentation*** of the(ir) art (Archer 2002: 88).[14] Smith (2010: 108) has commented that for Long, "the unrecoverable walk itself is his art" (for a discussion, see Morris 2020: 5)—and Long has expressed his belief that "[r]eally the only thing that reaches the art world is the image of the work" (quoted in "Interview with William Furlong (1984)" in Tufnell (2007: 60))—leaving Serota (2009: 38) to ask, "[w]hy do we accept a photograph of a sculpture made in the landscape as a surrogate for the experience of the whole walk?"[15] Writing in her catalogue essay to Long's exhibition at Tate Britain in London, Wallis (2009: 43) communicates that while photography has been "indispensable to Long's practice," his use of a camera "eschew[s] the technical mastery of the conventional fine art print. His photographs from the late 1960s are direct, rough and shot in black and white." Long would not disagree, explaining that he "always use[s] the same camera (Nikkormat) and the same lens. It is just to keep everything simple and straightforward. There is always the emphasis on the art and not really on the technique. It should look good but it should not look designed or special. The aesthetics should not get in front of the art" (in "Interview with Martina Giezen (1985–1986)" in Tufnell (2007: 78)).[16] For example, the "apparently artless" photograph taken by Long, *A Line Made by Walking* (1967) [see Fig. 1], "documents a walk back and forth across a park, which was repeated until eventually the grass along [Long's] route was trampled down" (Godfrey 1998: 130). "The meadow is an idyll," writes

Taylor (2009: 184), "except that Long then drains all sensuous content out of it with the meagre black-and-white photograph that embodies the work...."[17] As Godfrey (1998: 130) explains,

> it should be emphasized that it was the act of walking which constituted the work of art, although the photography has been exhibited and sold. The work is democratic (anyone could do it); it exists outside the circuit of the commercial gallery (at least until the gallery acquires the photograph); it is based on an idea—'walk back and forth until the grass is trodden into an evident line'—and it makes one think about one's relationship to the world. It is, of course, ecologically sound, since no lasting damage was done to the environment. If an artist normally draws a line, is walking a line a kind of drawing? Are we, then, all artists when we walk? In so far as Long was positing the possibility of being in nature, not just seeing it, but *being* in it, he is also making a political statement. We can escape from alienation and false consciousness: by a willful act we can be wholly in the world—reconnected.[18] [emphasis in original]

With this snapshot of the art and orientations of five artists (LeWitt, Weiner, Burden, Fulton and Long) in mind, let me now turn to some more general comments about Conceptual art and the relationship of the art to the artists' use of written declarations, descriptions, instructions, or statements, before attempting to draw comparisons to ecodefenders' use of the communiqué. As Godfrey (1998: 10) cautions, "we must... be wary of typologies, something which Conceptual artists have regarded as anathema." That said, Godfrey (1998: 7) offers the following:

> Generally speaking, [Conceptual art] may be in one of four forms: a *ready-made*, a term invented by Duchamp for an object from the outside world which is claimed or proposed as art, thus denying both the uniqueness of the art object and the necessity for the artist's hand; and *intervention*, in which some image, text or thing is placed in an unexpected context, thus drawing attention to that context: *eg* the museum or the street; *documentation*, where the actual work, concept or action, can only be presented by the evidence of notes, maps, charts or, most frequently, photographs; or *words*, where the concept, proposition or investigation is presented in the form of language. [emphasis in original]

Burden, Fulton, and Long's work—at least, that described above—involved photographic *documentation*, accompanied by *words* (and, in

Burden's case, "relics," as well), and thus I wish to highlight two aspects of the written features of/as documentation of Conceptual art.

First, the written declarations or documentary information provided by Burden, Fulton, and Long, as well as LeWitt, Weiner, and other Conceptual artists of the time, "were offered to audiences in lieu of sensual pleasures" (Hopkins 2000: 177) and, indeed, the printed material was treated as (or functioned as) an exhibition space (Rosenberg 2009: C5).[19] In so doing, not only did these Conceptual artists suggest that "[t]heory about art could be considered art" (Hopkins 2000: 179)—or, at least, some of them[20]—but they eliminated the traditional division between "the artist" (whose concern had been the production of a/the work) and "the critic" (whose role had been to evaluate and interpret). As Meyer (1972: VIII) describes, "[c]onceptual artists [took] over the role of the critic in terms of framing their own propositions, ideas, and concepts."

Second, just as these Conceptual artists raised questions regarding "what is (the) art?" and challenged traditional notions of art-as-material object (in the form of a painting or a sculpture)—just as they exploded notions of the temporal and geographic limits or confines of an exhibition and exhibition space—and just as they obfuscated, if not eviscerated, the line between artist and critic—those Conceptual artists using typographic statements (and/or photographs, maps, news clippings) redefined the role and possibilities of documentation. As Morgan (1994: 46, 47) explains,

> Traditional documentation of art objects provided information about art *without being a part of it*. The information was illustrative, descriptive or critical in relation to the artwork.
>
> . . .
>
> Prior to the advent of Conceptual Art, documentation was primarily considered *a secondary source which was removed from the art object*. This was because the art object 'housed' the form of the piece; and to perceive the form of the piece, one must observe the art object. In Conceptual Art, the secondary or external function of documentation changed. Instead of having to perceive the form of the piece through the art object, one could not 'see' the structure of the piece directly through the observation of documents. The documents, however, were only representative of the artist's ideas; therefore, to understand a work of Conceptual Art, one must reconstruct the piece through a proper context of receivership. [emphasis added]

In other words, prior to Conceptual art, the work of art—the art object—could exist without the documentation (written or otherwise); whatever documentation that did appear served as a supplement, perhaps revealing something of the artist's inspiration or intent or guiding the viewer toward a particular interpretation of or interaction with the piece. With Conceptual art, documentation (written or otherwise) still stood as a reference to something else/something other than itself, but that "something else"—the referent—was or became an *idea*, not (encapsulated in) an object with any material presence or permanence, such as a painting or sculpture, and without any aesthetic value.

With this background, I turn now to some shared aspects of Conceptual art and ecodefense and then suggest how a consideration of Conceptual art's (relationship to) written documentation can help reveal some of the meaning and purpose of ecodefense and its communiqués. Before doing so, however, it is necessary to mention a few disclaimers.

First, and as noted above, just as not all acts of ecodefense are accompanied by communiqués, not all Conceptual artists used/have used written declarations, descriptions, instructions or statements,[21] and many artists who rejected/have rejected formalism and who have sought/seek to promote ideas about art over art objects did/do not consider themselves "Conceptual artists" (see, e.g., Godfrey 1998: 153; Kennedy 2007: AR31; Meyer 1972: XIIn.16; Morgan 1994: 150). Second, just as the examples of ecodefense mentioned in the previous chapter involved arson and property destruction—a fraction of the range of actions/tactics available to the ecodefender—I have limited my analysis of Conceptual art to those artists who relied on words and documentation (in written form, in photographic form or some combination thereof) as "evidence." There were and continue to be, of course, other types of Conceptual art, as suggested by Godfrey's (1998: 7) "four forms," above (i.e., *readymade*, *intervention*, *documentation*, *words*). Indeed, as Morgan (1994: xiv) makes plain, "[e]ach work of Conceptual Art is different, contingent upon the intentional predilection of each artist, and the documents differ according to their specifications and context." Third, just as I focused on instances of ecodefense in the United States, I limited my discussion to Anglo-American Conceptual art. It is important to remember that ecodefense, as I have conceptualized it, has occurred worldwide (recall that E.L.F. formed in Brighton, England, in the early 1990s), and that "Conceptual art was, and is, a truly international phenomenon" (Godfrey

1998: 7)—as evidenced, in part, by the Queens Museum of Art's exhibition, *Global Conceptualism: Points of Origin, 1950s–1980s* (Camnitzer et al. 1999; see Vetrocq 1999; see also Bird and Newman 1999: 1; Hopkins 2000: 255).[22]

These disclaimers aside, both ecodefense and Conceptual art—at least, the art produced in the late 1960s and early 1970s—possess destructive traits or appetites for destruction stemming, in part, from an anti-capitalist orientation or spirit (see, e.g., n.[18] and accompanying text). As Eastwood (2006: 117, 121) argues, "capitalism requires that consumers *not* be concerned about the environmental degradation and social costs caused by corporations in the process of producing goods.... [O]ur consumerist society has environmental repercussions.... The fact remains that current extractive industries and our fossil fuel-based economy, which emphasizes capital gain as the primary value over other concerns (environmental, labor, or human and animal rights) has devastating impacts on the environment and on people who are either incorporated as resources into the amassing of capital or who are least able to afford to avoid environmental degradation" (emphasis in original). Ecodefense, as Eastwood (2006: 123) implies, such as the torching of fuel-inefficient vehicles and suburban development projects, represents, at least on some level, a challenge to global neoliberal capitalism and its concomitant negative environmental impacts (because, in her words, it "hits precisely at one of the values that Americans hold dear—private property").[23] While Conceptual art's destruction of artistic tradition was much more metaphorical,[24] the artists of the late 1960s and early 1970s working in a conceptual vein seem(ed) to share some of the motivations and worldviews of contemporary eco-defenders. Morgan (1994: 80, 104; 1996: 21) suggests that the Vietnam War caused many to question the American way of life, and, as such, he informs us, "[m]any of these Conceptualists turned to various forms of Marxism in order to sustain their purpose in de-emphasizing the art object in favor of nonmaterial concerns" (1994: 111 (footnote omitted)). While Godfrey (1998: 15, 190) points out that there was little *direct reference* to Vietnam War or the student riots in Paris in 1968, "Conceptual art was made, it seems, in times of crisis, when authority—both political and artistic—came into question.... [I]n its discomfort with the consumer society, Conceptual art was very much a reaction to and against this time." Admittedly, there is a difference between burning a fuel-inefficient vehicle and figuratively firebombing art history and aesthetic tradition,[25] but the common critique of social and cultural value systems (see Bird and Newman 1999: 2), is not insignificant.

In addition, some Conceptual artists seemed to share the ethos of contemporary "radical environmentalism." Fulton's and Long's walks were purposely envisioned as and deliberately intended to be gentle and ephemeral—consistent with the "leave no trace" principles of many ecodefenders—while Weiner ([1969] 1972: 217) was particularly pointed in his criticism of the permanent damage caused by humans to nature: "Industrial and socioeconomic machinery pollutes the environment and the day the artist feels obligated to muck it up further art should cease being made. If you can't make art without making a permanent imprint on the physical aspects of the world, then maybe art is not worth making."

While Conceptual artists and ecodefenders may share a certain political perspective and concern for the Earth, my interest here is not in comparing Conceptual art to ecodefense (or Conceptual artists to ecodefenders) but in considering the action-communiqué relationship of ecodefense to the art-written document relationship of Conceptual art. As suggested in this chapter, for many Conceptual artists, the "art" was the walk (in the case of Fulton and Long) or the shot/shooting or crucifixion (in the case of Burden). In other words, the "art" was the act(ion) or event and the exhibited text merely an ex post facto description of art that had already occurred. Similarly, for most ecodefenders, the act(ion) is the property destruction or sabotage or the blocking/defense/protection (or the liberation of live animals from laboratories) and the ex post facto communiqué, if issued, describes the action that has already transpired. That said—and here is where the analogy starts to break down—for many Conceptual artists, the activity or action described did not or need not be undertaken: the words could or can stand alone—or, as Weiner stated, "The piece need not be built." Not so for ecodefense and ecodefenders. While direct action without a communiqué is still direct action (although its purpose perhaps obscured and its pedagogical potential attenuated), "radical environmental" activists may consider words without direct action to be just words: nothing gets liberated, prevented, saved, or stopped.[26]

Despite this difference, Conceptual art can help illuminate certain features of ecodefense and thus **the act of ecodefense: communiqué:: work of conceptual art: written declaration/statement** analogy is not without utility. First, I would suggest that Conceptual art's written documents help differentiate other documentation from mundane objects and help ensure that these other forms of documentation *not* be considered aesthetically. As Morgan (1994: 82–83, xiv) explains, "[a] document generally functions as supportive evidence; it gives 'information' about something that has occurred.... Rather than presenting the viewer with a material object, a painting or sculpture, for instance, in traditional terms, the Conceptualists presented statements to be read, usually accompanied by documentation intended as supportive evidence of the concept or system." Thus, for example, although Burden did not consider the nails and written description to be (the) art (*Trans-fixed*)—the "art" having occurred on April 23, 1974—the written statement helped distinguish the nails from any old nails on a workman's bench (although the blood might raise some eyebrows) or, even worse, from a *sculpture* to be assessed and valued as a three-dimensional entity. In other words, if one were to remove the text and change the title from *Trans-fixed* to *Homage to Christ*, a viewer might perceive the nails as a reflection of the artist's religious faith and convictions. Similarly, with *Shoot*, the absence of text to accompany the photograph might lead one to dismiss the blurry black-and-white photograph as a beginning photography student's exercise in film speed, lighting and developing. The communiqué, I would argue, performs a similar function for an E.L.F. action—distinguishing it from anomic or hate-based arson—and thus transforming the event by providing an explanation, guiding interpretation, and shaping meaning.

Second, I would propose that the level and nature of engagement and (self-)reflection generated by Conceptual art, in general, and Conceptual art's written features, more specifically, bear a likeness to that brought about by ecodefense, as expressed in its communiqués. As Godfrey (1998: 4) explains,

> Conceptual art is not about forms or materials, but about ideas and meanings. It cannot be defined in terms of any medium or style, but rather by the way it questions what art is. In particular, Conceptual art challenges the traditional status of the art object as unique, collectable or saleable. Because the work does not take a traditional form it demands a more active response from the viewer, indeed it could be argued that the Conceptual work of art only truly exists in the viewer's mental participation.

This was—and is—especially true with Conceptual artists, such as Weiner, who, in Kennedy's (2007: AR31) words, "leave[s] the viewer with a lot of responsibility." While *all art* requires *something* of the viewer, "'[Weiner] makes that requirement a huge part of his art.... You really have to look and think in a way that demands more of you as a person and as a viewer. It doesn't coerce you. You have to engage it'" (Donna De Salvo, Chief Curator and Associate Director for Programs at the Whitney Museum of American Art, quoted in Kennedy 2007: AR31). Essentially, "Conceptual art concerned itself to a great degree with an examination of what art was: what were the necessary and sufficient characteristics required for a thing to be deemed art, and how might it be exhibited, curated and criticized?" (Archer 2002: 74). Or, as Godfrey (1998: 15) observes, "Conceptual art is concerned both with intellectual speculation and with the everyday. Conceptual art asks questions not only of the art object: 'Why is this art? Who is the artist? What is the context?'—but also of the person who looks at it or reads about it: 'Who are you? What do you represent?'"

In some ways, ecodefense, when accompanied by an explanatory communiqué, requires a similar type of engagement and introspection. With a terrorist activity—one in which bodily harm and indiscriminate killing for the purpose of instilling fear is intentional (see Chapter 1, n.9 and accompanying text; Malm 2021: 107; see also Malm in Illing 2021; Vanderheiden 2005: 428, 436; see generally Amster 2006: 300; Loadenthal 2017: 2–6)—we might struggle to disentangle complex and varied motivations (see, e.g., Bergen 2016), focusing our response on the myriad of reasons for what drove someone or some group to engage in an (arbitrary) attack to end human life; because "radical environmentalists," such as those participating in A.L.F. and E.L.F. actions, take precautions *against* harming human life—the raison d'être of terrorism—and because A.L.F. and E.L.F. have never caused human injury or death—when confronted by an incident of ecodefense, we cannot (or should not) attempt to analyze, consider, and understand the action in the same way as a terrorist act (see Illing 2021; Lennard 2019; Loadenthal 2017: 16, 17; Marchese 2024; Malm 2021: 155; Pickering 2007; Resistance Magazine 2010; Salter 2011; Smith 2008; see also Amster 2006: 288; Brown 2021: 365 (citing Brown 2020: 246); Farrer and Klein 2022: 219). In other words, because A.L.F. and E.L.F. tactics are "noninjurious by design" (Brown 2020: 240)—because the more risky forms of ecodefense, such as those acts of arson and property destruction described in the previous chapter, stop short of the ultimate act (the taking of another's life)—just as Conceptual art stops short of making a physical piece—we are presented with a range of questions similar to those suggested by Godfrey (1998) above. Instead of "Why is this art? Who is the artist? What is the context?" and "Who are you [the viewer]?" and "What do you [the viewer] represent?," we might ask, "Why is this individual or group of individuals attempting to defend the Earth in this way? Who are they? What is the context for this direct action?" and "Who are we in relationship to this action and to the world around us (e.g., victims, bystanders, perpetrators of environmental harm, less 'radical environmentalists')? What do we represent?"

In presenting this argument—in asserting that we should "look at the phenomena of the world in an 'art' kind of way" and ask "[s]uppos[e] I look at this as if it were art?" (Archer 2002: 88)—I want to make very clear that I do *not* consider acts of ecotage or monkeywrenching *to be art*.[27] While, as Hayward (2004: 67) maintains, "art is always important because it represents and reflects the essence of our cultural times," my suggestion here that we look at something *as we would look at art* does not mean that the action, event, phenomena, or thing *is art*. When the late German composer, Karlheinz Stockhausen, referred to the attacks of September 11, 2001, on the World Trade Center in New York as "the biggest work of art there has ever been.... the greatest work of art that is possible in the whole cosmos" (Gillen 2004: 131; Lentricchia and McAuliffe 2003: 6; Tommasini 2001), he was excoriated—and rightfully so—for his insensitivity and "reckless commentary" (Tommasini 2001).[28] As Anthony Tommasini, chief music critic for *The New York Times*, eloquently stated in response to Stockhausen, "[a]rt may be hard to define, but whatever art is, it's a step removed from reality. A theatrical depiction of suffering may be art; real suffering is not" (Tommasini 2001).[29] Ecodefense is not Conceptual art or any other kind of art, for that matter. But like Conceptual art—and art more generally—ecodefense is a complex, dynamic phenomenon, and my goal here has been to focus on and better understand one aspect of it: the relationship of the direct action to the communiqué. Recognizing, as Cottee and Hayward (2011: 964) do, that "there is a moral hazard involved in being *too* understanding [of terrorists]" (emphasis in original; endnote omitted)[30]—and thus that there may be some risks in trying to interpret and comprehend ecodefense—I assert that *not* attempting to make sense of acts of ecodefense is also risky for it obscures the range and specifics of environmental issues and problems, conflates the different and vibrant means of preventing or resisting environmental harm, and ignores the power struggle over the motives for and meanings sought and implied by particular actions (see generally Meikle 2002: 160).

When we experience a work of Conceptual art, we contemplate the artist's statements and documents as a manifestation of an idea and a record of (a) process. And we tend to judge the works in terms of the tightness of the relationship between the idea and this documentation or manifestation of it. We should, in turn, explore the communiqué-action relationship and assess the clarity of the statement in relationship to the action and question whether the action is the most effective means of achieving the stated goal of/in the communiqué, as well as the environmental impact of the action itself. (For instance, did the action, to quote Smith (2018: 22), "disrupt[] practices that might result in or immediately threated to generate, serious and irreversible harm"? Did the action cause pollution or loss of nonhuman life that exceeded the damage that had or would have occurred from the project or activities?[31]) We should take into account whether and to what extent the communiqué states and explains what transpired, threatens or warns present and future targets, attempts to raise awareness of a specific and/or pervasive environmental problem, and/or offers alternatives and guidance for how to live differently/more harmoniously.

Notes

1. Such as with the image on the cover of this book adapted from a beautiful oil painting, *Controlled Burn* (2024), by the Lexington, Kentucky-based artist, Lina Tharsing.
2. Moorhouse (2002: 32–33) proclaims that "Duchamp's designation of pre-fabricated objects as works of art is the source of a major tributary running through Modernism." But this, as Moorhouse (2002: 33) avers, has "limited relevance to an appreciation of [Richard] Long's designation of walking as art"—discussed elsewhere in this chapter. "Duchamp's innovation—bold as it was—rests on the perception of one or more physical objects," Moorhouse (2002: 33) continues. "In contrast," Moorhouse (2002: 33) continues, "Long took sculpture into the domain of the immaterial: the walk comprises the movement of a body through time and space and, as such, has no permanent physical attributes." Moorhouse (2002: 33) insists—although I disagree—that "the walk as artwork [cannot] be understood by reference to avant-garde art

movements such as Conceptual Art, whose emergence formed the backdrop to Long's own artistic development. The premium placed by conceptual artists on ideas—to the extent that their actual realisation is secondary—is completely antithetical to Long.... the walk itself may have no lasting physical attributes but the work could not exist if the walk had not happened." Later in his essay, Moorhouse (2002: 36) doubles down on his commitment to this stance:

> [t]he relationship between the idea for a walk, the walk itself, and the physical evidence of the walk, is a fundamental issue in Long's art. While it is possible to identify these different components singly, it is the interaction of these components that provides the fabric of his work. It is for this reason that, as already seen, Long's art falls outside the definition of Conceptual art. The idea is vital in that it defines the structure for a walk. But the walk is equally important in that it realises the idea, actualising the structure as physical movement through time and space, so that the work of art has a *real*—if transient—existence. [emphasis in original]

What Moorhouse (2002) does not appreciate is that only *some* Conceptual artists subscribed to the belief that the idea did not need to be "realised" (at all). (Lawrence Weiner, also discussed elsewhere in this chapter, serves as a good example.) What renders Long a *Conceptual artist*, at least in part, and his walks *Conceptual art*, among other things, is his elevation and championing of ideas over art *objects*, his abjuration of concerns with artistic skill or craft, and his passion for impermanence—not, as Moorhouse (2002) would have us believe, a rejection of (the) actualization (of ideas). The notion of "realisation as secondary" may have been a *hallmark* of (some) Conceptual art, but it was not *the singular requirement*. Admittedly, Long ([1988] 1991: 251) has asserted that "[i]t is no good just having a good idea, it is also necessary for me to make it, and also not have somebody else make it [as Sol LeWitt might]—for me to do it myself, because my work is my own footsteps, it is only what I can do"—a position he has expressed in other contexts:

> It's . . . very important that my work is made by me, and I don't mean that in an egotistical way, but a fundamental one. . . . [I]n my work, the centre of my work is that the walking is done by me. . . . And that's partly because it's a pleasure to make it. For

me, the emotional pleasure of being an artist and making my work is actually doing it. I guess, as you say, the part to the whole is also measuring myself against the landscape. If I walk across the country it's however many days it takes *me* and *not somebody else*. So it's always particular to me; it always comes down to me [sic] doing it. I'm the measure against the place.

[quoted in "Interview with Michael Auping (2000)" in Tufnell 2007: 115–116; emphasis added]

And:

At the heart of my work is the fact that I make it myself. I have the pleasure of walking and being in these places and making these works in the landscape. It's also a question of realizing particular ideas, too. So let's say I have a certain idea about the symmetry of time and places, I can actually carry it out personally by walking that idea. In other words, there's no point in getting somebody else to do it. The point of my work is my own physical engagement with the world in different ways . . . .

[quoted in Ayers 2006; O'Rourke 2013: 266 n.75]

That being said—and unlike Hamish Fulton (see Collier and Morrison-Bell 2013: 63)—Long has acknowledged a connection with or tie to Conceptual art (see, e.g., Ayers 2006; Long [2008] 2009: 172; Wallis 2009: 35; cf. "Interview with William Furlong (1984)" in Tufnell 2007: 90; Seymour 1991: 8, 12), thereby calling into question Moorhouse's (2002: 33) dogged conviction that Long's "walk[s] as artwork [cannot] be understood by reference to avant-garde art movements such as Conceptual Art." For a discussion of Richard Long's *Turf Circle* (1966)—which "contains a smaller central circle or axle that turns it into a prehistoric bicycle wheel"—as an "homage… to the oblique conceptualist aesthetics of Marcel Duchamp," see Taylor (2009: 181 (endnote omitted)). For Fuchs (1986: 101, 72), while Long's art "shows no inclination towards artistic *complication* of form" (emphasis added), it still does possess "a very strong basic formal component which is not alien to typical Modernist concerns in contemporary art." For Long's thoughts on Weiner's "objectless art," see Wallis (2009: 48); see also "Interview with William Furlong (1984)" in Tufnell (2007: 90), where Long states, "Lawrence [Weiner] hates the label 'Conceptual Art' just as I hate 'Land Art', but I think he is the most important artist from that so-called Conceptual generation. The discovery he made that art does not necessarily have to be made, that was a great breakthrough."

3. Kennedy (2007: AR31); see also Alberro and Zimmerman (1998: 50); Archer (2002: 75); Batchelor (1998: 76); Chiong (2007: 325); De Salvo (2007: 65); Gillick (2007: 244, 250); Goldstein (2007: 110); Rorimer (1999: 13); Weiner (1972: 218). Archer (2002: 75) adds that "Weiner refused to make assumptions about the viewer, always accompanying his texts with [this] short statement."
4. Godfrey's (1998: 166) phrasing is a little confusing. He writes that "Weiner's 1968 exhibition at the New York gallery run by Seth Siegelaub consisted of twenty-eight phrases, which were published, one per page, as a book entitled *Statements*." Phillpot (1998: 35), however, asserts that "Siegelaub had no permanent gallery space at the time," and that Weiner's *Statements*, "described as his 'third show with Seth Siegelaub'" (quoting Schwarz 1989: 136), was, in fact, *a book*—"simply a collection of phrases, such as 'One standard dye marker thrown into the sea,' which [could] be regarded either as descriptions of works or as prescriptions for future works."
5. Reprinted in Godfrey (1998: 166). Goldstein (2007: 123) explains that "[t]he statements are not only materially present in language, but also 'built' in their typeset form, which gives them an object status—each statement is formatted into narrow columns, many of which are broken mid-word." Or, as Weiner (1998: 22) stated, "Just because I use language, it does not give me the inherent responsibility to be a grammarian, or a linguist."
6. For a discussion of *Trans-fixed*, see, e.g., *BBC News* (2017); Boucher (2017); Fox (2015); Jaworowski (2017); Kaye (2018: 27, 28, 29, 30, 41, 49); Kennedy (2013); for an image of Burden crucified on the back of the Volkswagen Beetle, see, e.g., *BBC News* (2017); Boucher (2017); Jaworowski (2017); Kaye (2018: 28); for a photograph of the "relics" that were exhibited, see Kaye (2018: 29).
7. Kaye (2018: 41) does not separate the written description and nails from the performance—or, shall I say, he does not isolate the physical objects (the evidence of the performance) so completely from the performance, itself, contending that "the shock in encountering the nails as relics of Burden's *Trans-fixed*... are *entangled*

with Burden's live act" (emphasis added). While Kaye (2018: 30) acknowledges the "contextual form, meaning and purpose of *Trans-fixed*," he maintains that "'its' meanings cannot be divorced from the time of its execution and those which accrue"—***and have accrued***, I would add—"subsequently in relation to that occasion." For Kaye (2018: 49), "[t]he 'original' act of *Trans-fixed* was never 'in possession' of its meanings, which flowed from its dissemination and representation: its meanings did not coalesce *within* a discrete performance now lost… but in its subsequent circulation over time, its *ramifications*" (emphasis in original). In other words, Kaye (2018) appears to be suggesting that the meaning of *Trans-fixed* was not something that could be appreciated or understood *only* by whoever happened to witness Burden nailed to a Volkswagen on April 23, 1974. Rather, Kaye (2018) implies that the "'original' act"—what took place on April 23, 1974—contained *a* meaning (or *multiple* meaning*s*), but that *other meanings* (have) emerged over time (as people (have) experienced the "relics" and image as documentation) and that these cannot be ignored, downplayed, or dismissed.

Kaye's (2018) perspective would seem to be at odds with Conceptual art's insistence on the "dematerialization of the art object" (Lippard [1973] 1997), where *the idea* is *the work* (and thus paramount) and a/the material form is secondary (because it is documentation—simply recording the idea or process). But in many respects, Kaye may really just be pointing out that a work of art has a context in which it is made and that *that* context is important to/for its meaning—and, perhaps, *just as vital* in/for performance art (which is how classifies *Trans-fixed*) or Conceptual art than in/for, say, a painting made at a specific time and place (such as *Guernica* (1937), which the Spanish artist Pablo Picasso painted in his home in Paris in response to the April 26, 1937, bombing of the Basque town of Guernica). Essentially, just as a painting's (initial) meaning and significance may be linked to a certain time, but take on new weight and importance later (see generally Brisman 2011: 16-17n.5)—*Guernica*, for example, remains a powerful and moving anti-war symbol more than eighty years after it was created (see generally Hoberman 2004)—a work of performance art or Conceptual art can hold (different) meaning(s) and significance long after the performance or "'original' act" despite (or perhaps *because of*) the fact that the documentation is all that remains—and despite whatever the artist may claim is actually the (real) art.

8. The marksman was Bruce Dunlap, a fellow art student of Burden's, whose identity was kept secret for many years (see *BBC News* 2017; Kutner 2015; see also Boucher 2017; Ghorashi 2015).
9. The *BBC News* (2017) shares this perspective, stating that "Burden's idea was that the act of doing something was the art, a concept which also meant the artwork couldn't be bought or sold. He subsequently suggested that these active works were an attempt to control fate, or at least give the illusion that such a thing was possible." Bleeker (2018: 251), on the other hand, insists that "documentation is co-constitutive of what will become the performance, even in the case of performance documentation that is supposedly a recording of an event that took place independently of the act of recording, as, for example, Chris Burden's action *Shoot* (1971)." Citing Auslander (2006),Bleeker (2018: 251) adds that "even such archetypal works of body art and performance as Burden's were not autonomous performances, the documentation of which supplements and provides access to an originary event, but staged to be documented at least as much as to be seen by a live audience."
10. Serota (2009:29) observes that "Long's own body [is]... the instrument of his work. The body has always been implicit with its weight and passage expressed through the trodden grass in *A Line Made by Walking* 1967 or the straight path beaten in the dust of the Peruvian desert in *Walkinga Line in Peru* 1972." On the relationship of Fulton's "walking body and the landscape," see Morris (2020: 60).

    Note that Fulton and Long have taken walks together (see, e.g., Fulton 1991; Long [2008] 2009; O'Hagan 2009; Taylor 2009: 182; see also Collier and Morrison-Bell 2013: 63; Fuchs 1986: 73; "Interview with William Furlong (1984)" in Tufnell 2007: 59–60; "Interview with William Furlong (1984)" in Tufnell 2007: 68; "Interview with Colin Kirkpatrick (1994)" in Tufnell 2007: 100). For a description of early collaborations between Fulton and Long, see Wallis (2009: 42–43).
11. Morris (2020: 2) in his discussion of "Contemporary Considerations of Walking as an Artistic Medium"—the first section of

his chapter, "Walking: A Distinct Artistic Medium" in *Walking Networks: The Development of an Artistic Medium*—signals a contrasting perspective, pointing to the exhibition catalogue for *Walk-On: From Richard Long to Janet Cardiff—40 years of Art Walking* (2013), in which curator Cynthia Morrison-Bell "identifies a difference between """walking as art," and "art walks" themselves'" (endnote omitted). "In the first category," Morris (2020: 2) summarizes, "are the sculptures of Richard Long, the textual works of Hamish Fulton, and the video art of Francis Alÿs; the second, smaller category consists of the audio walks of Janet Cardiff and George Bures Miller, or Fulton's choreographed group walks."

12. In a press release for a solo exhibition at the Royal West of England Academy in Bristol in 2000, Long specifies that his "walks are recorded or described in [his] work in three ways: in maps, photographs or text works, whichever form is the most appropriate for each different idea" (in Tufnell 2007: 39; for a similar sentiment, see Long in "Interview with William Furlong (1984)" in Tufnell 2007: 69; Long in "Interview with William Furlong (1984)" in Tufnell 2007: 105). Macfarlane (2009) remarks that "[w]hen questioned about the textworks, he [Long] plays an amusingly straight bat: '"Heathrow Airport" means I just passed Heathrow Airport. "Dead Stoat" means I passed a dead stoat in the gutter'" (see also "Interview with William Furlong (1984)" in Tufnell 2007: 75). Moorhouse (2002: 38) explains that "[i]n Long's work, maps, photographs and words relate to a walk or sculpture as a footprint does to a foot. As signs, they are partial, revealing aspects of the experience that caused them; but they also provide—in their different ways—sufficient information for the viewer to *imagine* the circumstances which led to their creation" (emphasis in original). Moorhouse (2002: 38–39) clarifies, though, that "[s]ince 1977, Long has used words less as factual or descriptive annotations to photographs or maps and more as a discrete way of working in its own right. The role of the text works is to tell the story of individual works. They take the form of words or phrases—notable observations, feelings, experiences, place-names, measurements of time, duration, number and distance—drawn into particular arrangements." But they are "rarely constructed as narratives," Moorhouse (2002: 39) continues: "Neither poetry

nor straightforward prose, the structure of such texts is closer to sculpture than to literature, arising from the connection and interrelation of words, ideas and experiences."

13. Higgins (2012) quotes Long for a similar statement: "There is a point of view... that if you go into the landscape you should only leave footprints and take photographs. The other extreme is making monuments. I have no interest in making monuments. But I think there is a fascinating territory between those two positions. I can move things from place to place. I can manipulate the world by leaving stones on the road. And they don't disappear because the stone is still in the world—but completely anonymously."

    The artist, Michael Craig-Martin, in a conversation with Long, contemplates how Long's work "in the landscape . . . [is] never about domination, it's always about trying to identify with the place. . . . I've always thought of American artists' attitudes to the land as typically American—a frontier spirit, the need to overcome nature in order to survive. [Long's] approach has always been to go *with nature*" (in Long [2008] 2009: 173 (emphasis added)). Seymour (2002: 9) makes similar observations about Long's work and careful approach. In her essay in *Richard Long: Walking in Circles* (1991: 12), Seymour explains that "[u]nlike certain American 'land artists', [Long] does not shift great quantities of earth or make permanent monumental works," and in the Foreword to *Richard Long: Walking the Line* (2002: 9), muses:

    > Our world is made up of lines, from comet tails to DNA. Everything is connected. Everything is sequential. Everything that moves, from a snail to a lava flow, leaves a line, a trace of its passing. A line can be fate, a commitment, a fact, a relationship, a place. Some lines are well trodden paths, some intersect, some pass at a distance, some return to their origins. We all walk the line. We have an end and a beginning which is joined to a much longer invisible line in the past and in the future.

    > Long's touch has always been famously light. Unlike some land artists, he has never made permanent changes to the land or moved great quantities of earth.

    This last point by Seymour (2002) is helpful in further distinguishing Long from some of his contemporaries. As Macfarlane

(2009) claims, Long's "early work was categorically different to the land art projects that were under way in America in the same period (Robert Smithson, Hans Haacke, James Turrell)"—a statement that resonates with Craig-Martin's reflection above (although one should note that Haacke is German-born). "Long's interventions were more modest," Macfarlane (2009) continues, and "his sculptures bigger but less massy. The same year that Smithson was hiring earth-movers to bulldoze his Spiral Jetty into place on the salt flats of Utah (a 1500ft chameleon's tongue of black basalt, curling out into ruddy water), Long was walking due north over Dartmoor. As Turrell was starting to reshape an extinct volcanic cinder crater in Arizona, Long was arranging a small circle of stones in the Andes."

In a related vein, Jones (2009a) gushes that "wherever he goes, Long makes a mark," and, "in that sense," Jones (2009a) maintains, "he's an inveterate humanist—we must mark our passing, it's in our nature." But, Jones (2009a) concludes, Long "shows us that the simplest, most impermanent and sympathetic marks can be the most eloquent." This seeming contradiction—that Long's work "makes a mark" but is also "impermanent" or evanescent—is encapsulated nicely by Serota (2009: 28, 29):

> Long has chosen some awesome places both to traverse and in which to leave a mark, but he has also elevated the significance of places and sights that we all take for granted: the crossing place of road and river, the passing of a bird in flight. His observations of the weather, animals and plants encountered on a journey have become a more insistent strain in his work. Occasionally there is an oblique reference to other human activity, but usually Long captures the experience of being alone in a world governed by the passage of the sun and the cycles of nature.
>
> . . .
>
> His works are time in their classic rhythm and beauty, but also particular to this moment. Few artists make us more aware of the power and the fragility of the earth or, indeed, our own brief passage across its face.

It is not surprising that "walking artists" would eschew connections to "land art" or the "land art" label. Long has stated, "I hate 'Land Art'" (in "Interview with William Furlong (1984)" in

Tufnell 2007: 90), while Fulton, the contemporary artist whose work shares the greatest affinity with that of Long, has proclaimed that he "want[s] no association with Land Art" (quoted in Collier and Morrison-Bell 2013: 63).

14. Lorimer (2016: 26) explains that "[t]he original walk demands no audience, is undertaken alone, designed to leave no lasting trace of passage, and other than having been completed may well end with no discernible 'product'." Of Long, Fuchs (1986: 134) reminds us that many of his works are or were "made in isolation, one to one with the place, and the spirit of the work should not be altered by visitors. That is why it is not important whether outside works remain or change or disappear. Information about them can be transmitted through words or photography. But words, and especially photography, have a tendency to enhance distance and remoteness: what you see in a photograph is in past time and somewhere else." Note, however, that since his first works, Fulton has undertaken what has been referred to as "group walks" (Morris 2020: 57–61; see also Collier 2013: 75) and "communal walks" (Collier and Morrison-Bell 2013: 63).
15. Over the years, Long's approach or perspective as to what constitutes "the art" seems to have softened. Wallis (2009: 46, 47) suggests that "[o]ver time... Long realised that a photograph could be a work in its own right, not the sculpture or walk itself, but the work represented or put into the public domain in another way.... The reproductibility of the photograph facilitated the existence of different versions of the work and... Long realised there were different ways to present his art...." For a similar sentiment, see Long in "Interview with William Furlong (1984)" in Tufnell (2007: 61); Long in "Interview with William Furlong (1984)" in (Tufnell 2007: 65); Wilson (2009).

    In a slightly different vein, Long ([2008] 2009: 177) remarks, "I often think—if I had just made *A Line Made by Walking* it would have been a good work in its own right. But the fact that I've repeated or continued that idea for many years, walking lines in different ways, for different reasons, in different landscapes, that's part of the life of *that* work. The cumulative effect of all the other works consolidates and informs that first idea . . . ." (emphasis added). (Long reflects on remaking *A Line Made by Walking* (1967) in 1983 for *Omnibus*, the British television program on

BBC TV, in "Interview with William Furlong (1984)" in Tufnell (2007: 67), and he adds that many of his subsequent (longer) walks have constituted "an extension" of *A Line Made by Walking* (1967) (Tufnell 2007: 68).)

16. Later, in the same interview, Long ruminates on the difference between "young artists" and artists of his generation: "Now the most exciting thing is what it actually looks like. You know, how a painting looks. So, as well as a return to object making, painting and figuration, mythology and so on, it is a return to aesthetics. Whereas for my generation *the ideas were the most important* and then the aesthetics looked after themselves" (quoted in Interview with William Furlong (1984)" in Tufnell 2007: 81).

    Regarding his use of photography, Long ([1988] 1991: 252) feels that "[e]ven though it is necessary to get a good photograph, the photographs should be as simple as possible so that when people look at the photograph they are not dazzled by wide-angled lenses or special effects" (see also Long in "Interview with William Furlong (1984)" in Tufnell 2007: 61; Long in "Interview with William Furlong (1984)" in Tufnell 2007: 65). "Because my art is very simple and straightforward," Long ([1988] 1991: 252) explains, "I think the photographs have got to be fairly simple and straightforward . . . ."—and elsewhere, he refers to his photographs and their captions as merely or simply "facts" (see Long, "Five, Six, Pick up Sticks, Seven, Eight, Lay Them Straight" (1980), in Fuchs 1986: 25–27 and Tufnell 2007: 15; Long, "Words After The Fact (1982)," in Fuchs 1986: 236 and Tufnell 2007: 25–27; Taylor 2009: 188n.6; see also Long in "Interview with William Furlong (1984)" in Tufnell 2007: 65). Distinguishing his approach to photography from that of Fulton, Long confesses, "I'm not really a great photographer, sometimes my photographs almost by chance are quite beautiful, whereas Hamish Fulton is a nature photographer. He has a great feeling for landscape images. The camera is a natural part of his body of a walk" (in "Interview with William Furlong (1984)" in Tufnell 2007: 90).

    For a thoughtful reflection on Long's use of photography, see Robinson (2013: 18); for some specific examples, see Phillpot (1998: 46).

17. Taylor (2009: 183) explains that Long, "[t]aking a bicycle from St Martins in central London to the southern edge of the city,...

selected a meadow and walked backwards and forwards across it in a straight line, flattening and rubbing the grass until the trace of his presence appeared visibly in the ground." Macfarlane (2009), who claims that *A Line Made by Walking* is Long's "best-known early piece," offers a slightly different description, recounting how, "[o]n a sunlit day in 1967, [Long] caught a train south-west out of Waterloo. When the suburbs gave way to countryside, Long got off the train, and found a field whose grass was starred with daisies. He walked back and forth, until the flattened grass caught the light such that it was 'visible as a line'. Then he photographed the line in black and white, and went home." (Wallis (2009: 43, 46) offers a similar version.) Moorhouse (2002: 37) reflects on *A Line Made by Walking*, stating:

> In keeping with his ethos of using means which are direct, immediate and practical, this took the form of a straight line running across a grass field: the result of repeating his steps. He then photographed the work. Simple to make and easy to describe, the implications of this piece were nevertheless profound. Essentially, its subject is the interaction of man and nature. Yet the way the work conveys this is economical, pure and appropriate. Without describing nature or adding anything superfluous, the work takes it place, temporarily in nature—its subject and its means brought into perfect alignment. The principle of making a work of art about the meeting of man and nature by leaving a sign of that interaction is encapsulated in this work and in Long's subsequent development of that approach.

18. With respect to Godfrey's (1998: 130) statement that *A Line Made by Walking* (1967) is "ecologically sound, since no lasting damage was done to the environment," Fuchs (1986: 70) would agree, noting that it "hardly disturbed the grass." (Long, himself, points out that *A Line Made by Walking* (1967) is "really a line which is visible because the grass is trodden down by my footsteps and the flattened grass reflects the sunlight, so *the sunlight is actually making the mark*" (in "Interview with William Furlong (1984)" in Tufnell 2007: 62).) Regarding Godfrey's (1998: 130) comments about "alienation" and "false consciousness," consider Smith's (2010: 108) observation that "[i]n 1987 'walking artist' Richard Long carried a stone from Aldeburgh beach on the east coast of England to Aberystwyth on the west coast of Wales, then

carried another stone back to his starting point (*Crossing Stones*, 1987); a banal kind of balance, an absolute reciprocity, reifying the walk as a settling of accounts in a dead economy, a parody of barter or exchange." As Smith (2010: 108) explains, "Long's work has been an explicit attempt to work outside the economy of everyday accumulation, dematerializing industrialized sculpture, deploying the anachronism of walking as a self-consciously 'primal art.'" To this, Smith (2010: 108) adds, that while Long's walks "may be archaic disruptions,... their documentation in high resolution, super-textured photographs and elegant texts, is retailed through the gallery system. The exchange value of these prints is raised by the inimitable strictures of the walks."

O'Rourke (2013: 43) considers Long to be "the first contemporary artist to see walking as an art form." For an overview of walking as an artistic medium, see, e.g., Robinson (2013); for an in-depth consideration, see, e.g., Morris (2020). On the relationship between walking, politics, and art, in general, see, e.g., Morris (2020: 7, 13-14, 37, 58, 94, 118, 163-164); on Fulton's "group walks" and interest in politics, see, e.g., Morris (2020: 7, 57-61).

19. Indeed, Phillpot (1998: 34) recounts how "[t]he significant event in terms of the history of artists' publications was [Seth] Siegelaub's Douglas Huebler exhibition in November 1968." "[F]or the first time," Phillpot (1998: 34) describes, "the exhibition appeared principally as a catalogue. The nature of Huebler's work was such that it functioned equally well on the page or on the wall, so viewing (or owning) the 'original' work, became, in effect, beside the point." Philllpot (1998: 35) continues: "At about the same time, Siegelaub (along with John W. Wendler), published *Carl Andre, Robert Barry, Douglas Huebler, Joseph Kosuth, Sol LeWitt, Robert Morris, Lawrence Weiner* (no. 92), an offset publication generally known as 'The Xerox Book' since it derived from photocopies made by the artists listed in its title. The printed pages in the book were not conventional art reproductions, though; they *were* the artworks. Thus, the book was a kind of collaborative artwork or anthology" (emphasis in original; endnote omitted).
20. Seymour (1991: 16) insists that "[a]lthough [Long is] firmly uninterested in theory, [he] seems to have a remarkable grasp of the whole language of art, perhaps because he regards it quite rightly as a natural phenomenon."

21. It is worth underscoring that Conceptual artists who (have) produced written declarations, descriptions, instructions, or statements do/did consider these to be *explanations* of their works. As Long ([1988] 1991: 92) asserts unequivocally, “my works should be completely self-contained, they shouldn’t need any explanations or references, they should be things or ideas in their own right.”
22. The exhibition, which was organized by the Queens Museum of Art (now known as just the “Queens Museum”), appeared at this museum in Queens, New York, from April 28, 1999–August 29, 1999. From there, it traveled to the Walker Art Center, Minneapolis, Minnesota (December 19, 1999-March 5, 2000), and then to the Miami Art Museum, Miami, Florida (September 15, 2000–November 26, 2000).
23. Echoing Eastwood (2006), Salter (2011: 231) observes that “actions that fall under the banner of ALF (and ELF) directly challenge the legitimacy of capitalism, in that property status of animals is rejected (rather than a focus on the treatment, such as those of reform-based organizations)” (footnote omitted).
24. Note, however, that one of Weiner’s early works—*Cratering Piece* (1960)—involved setting off a series of TNT charges (without permission) on the grounds of a state park in Mill Valley, California—explosions that “literally ‘carved’ out the earth” (De Salvo 2007: 61; see also Weiner 1998: 8; for a discussion, see Herbert 2021). For Weiner’s perspectives on art as/and metaphor, see, e.g., Goldstein (2007: 103); Leffingwell (2007: 308); Schwarz (2007: 178, 180, 185, 190).
25. That Weiner considered language a sculptural material (see Goldstein 2007: 102, 130; Weiner 1998: 12, 23) does not diminish the overall anti-formalism of Conceptual art.
26. This raises an interesting possibility: whether ecodefenders might contemplate issuing communiqués of direct actions *not* taken much like Weiner and LeWitt would offer descriptions or instructions of works not made. Of course, a communiqué untethered to an action would no longer be *direct* action—and depending on the forum and nature of the writing, it could range from *indirect* action to criminal threat. That being said, such an endeavor—a communiqué of an action not taken—would resonate with Weiner’s proposition that art need not be fabricated or performed but could or can be communicated through language alone or with

LeWitt's assertion that art could be communicated by language and realized by others (see Goldstein 2007: 104, 130; Hopkins 2000: 177; Strick and Weinberg 2007: 39). While perhaps interesting from an artistic perspective, my proposal for "a communiqué of an action not taken" would likely not be met with much enthusiasm by environmental activists—or even scholars of resistance, for that matter. As Hayward and Schuilenberg (2014: 27) have argued, there is "a danger that, by stressing creativity over and above other elements—say, for example, robust street-level protest and intervention—certain resistance activities might be reduced to *simple acts of communication*, empty cultural messages devoid of any real transformative potential" (emphasis in original). It is, as they argue, "important... for resistance to go beyond communication" (Hayward and Schuilenberg 2014: 28).

27. Nor do I view the communiqués as art, although I concur with Stephens-Griffin's (2023: 257, 260, 269) description of them as "cultural products" and "creative constructs."
28. Note, however, that after the fact, Stockhausen (2001) referred to the attacks of September 11, 2001, as an "atrocity," claiming that his statements were taken "out of context," and apologized if his "remarks were misconstrued to offend the grieving families of the brutal terrorist attacks on New York City and Washington D.C."
29. On the theatricality and performative elements of direct action, see Jordan (1998: 132).
30. As I have noted elsewhere (Brisman 2022: 463n.11), people may commit atrocities not because of a failure or lack of empathy but as a direct consequence of *over-identification* and a desire to *increase* empathy; well-meaning compassion can have or produce unintended consequences, such as intensifying conflict(s) rather than resolving tensions (citing Breithaupt 2019; Lambert 2019; Rosin 2019).
31. Smith (2018: 18), in putting forth his conception of an "ethic of responsibility," argues that "coercive campaigns should be *constrained*, *proportionate* and *discriminating*, such that their intended impacts are concentrated as far as possible on wrongdoing or wrongdoers" (emphasis added). While Smith (2018: 19) acknowledges that "there is a presumption in favour of coercive tactics that fall closer to the nonviolent than the violent end of the spectrum.... [t]here are... circumstances where certain forms

of property damage can be defensible, at least if the case for its use is sufficiently compelling. This is more likely to be the case if the property damage is proportional to the wrongs opposed by activists, carries no risk of physical or psychological harm to persons, and is strictly necessary to the goals of the direct action."

References

Alberro, Alexander, and Alice Zimmerman. 1998. Survey: Not How it Should Were it to be Built But How it Could Were it to be Built. In *Lawrence Weiner*, 34–74. London and New York: Phaidon.

Amster, Randall. 2006. Perspectives on Ecoterrorism: Catalysts, Conflations, and Casualties. *Contemporary Justice Review*, 9(3): 287–301. https://doi.org/10.1080/10282580600827991.

Archer, Michael. 2002. *Art Since 1960*, 2nd edn. London: Thames & Hudson.

Arn, Jackson. 2023. The Man Who Hurt Himself for Art. *The New Yorker*, March 31. https://www.newyorker.com/culture/the-art-world/the-man-who-hurt-himself-for-art-chris-burden.

Auslander, Philip. 2006. The Performativity of Performance Documentation. *PAJ: A Journal of Performance and Art*, 28(3) [84] [September]: 1–10. https://doi.org/10.1162/pajj.2006.28.3.1; https://direct.mit.edu/pajj/issue/28/3%20(84).

Ayers, Robert. 2006. The AI Interview: Richard Long. *Artinfo.com*, July 5. https://farticulate.wordpress.com/2011/01/29/29-january-2011-post-richard-long-selected-works-interview/ (last accessed October 26, 2024; on file with author).

Batchelor, David. 1998. Focus: Many Colored Objects Placed Side by Side to Form a Row of Many Colored Objects. In *Lawrence Weiner*, 74–83. London and New York: Phaidon.

BBC News. 2017. Shot and crucified: The extreme art of Chris Burden. *BBC News*, May 3. https://www.bbc.co.uk/programmes/articles/2S05hw48t0tpcnS03yqvrfp/shot-and-crucified-the-extreme-art-of-chris-burden.

Bergen, Peter. 2016. Why Do Terrorists Commit Terrorism? *The New York Times* (New York edition), June 15: A25. Published online as "Why Do Terrorists Commit Terrorism?" on June 14, 2016, https://www.nytimes.com/2016/06/15/opinion/why-do-terrorists-commit-terrorism.html.

Bird, Jon, and Michael Newman. 1999. Introduction. In *Rewriting Conceptual Art*, ed. Michael Newman and Jon Bird, 1–10. London: Reaktion Books.

Bleeker, Maaike. 2018. Resistance to Representation and the Fabrication of Truth: Performance as Thought-Apparatus. In *Artists in the Archive: Creative and Curatorial Engagements with Documents of Art and Performance*, Paul Clarke, Simon Jones, Nick Kaye, and Johanna Linsley, ed. pp. 247–269. Abingdon, Oxon, UK, and New York: Routledge.

Boucher, Brian. 2017. New Chris Burden Documentary Shows a Pioneer of Pain Who Drove His Loved Ones Crazy. *Artnet.com*, April 11. https://news.artnet.com/art-world/documentary-performance-legend-chris-burden-rebel-921092.

Breithaupt, Fritz. 2019. *The Dark Sides of Empathy*. Andrew B.B. Hamitlon, trans. Ithaca, NY, and London: Cornell University Press.

Brisman, Avi. 2011. Vandalizing Meaning, Stealing Memory: Artistic, Cultural, and Theoretical Implications of Crime in Galleries and Museums. *Critical Criminology: An International Journal*, 19(1): 15–28. https://doi.org/10.1007/s10612-010-9104-x.

Brisman, Avi. 2022. Immunity to Environmental Crime, Harm and Violence: An Ongoing Pandemic and a Possible Narrative Vaccine. *International Journal of Offender Therapy and Comparative Criminology*, 66(4) [March]: 451–469. https://doi.org/10.1177/0306624X20970885.

Brown, Joseph M. 2020. Notes to the Underground: Credit Claiming and Organizing in the Earth Liberation Front. *Terrorism and Political Violence*, 32(2): 237–256. https://doi.org/10.1080/09546553.2017.1364637. Published online on September 7, 2017.

Brown, Joseph M. 2021. Civil Disobedience, Sabotage, and Violence in US Environmental Activism. In *The Oxford Handbook of Comparative Environmental Politics*, Jeannie Sowers, Stacy D. VanDeveer, and Erika Weinthal, ed. pp. 356–374. Online edition, Oxford Academic, https://doi.org/10.1093/oxfordhb/9780197515037.013.34.

Brown, Mark. 2009. Art of the outdoors goes on show at the Tate. *The Guardian*, June 1. https://www.theguardian.com/artanddesign/2009/jun/01/richard-long-exhibition-tate-britain.

Burden, Chris. 1974. *Chris Burden 71–73*. Los Angeles, CA: Chris Burden.

Camnitzer, Luis, Jane Farver, and Rachel Weiss. 1999. *Global Conceptualism: Points of Origin, 1950s–1980s*. Queens, NY: Queens Museum of Art.

Chiong, Kathryn. 2007. Sympathy for Lawrence Weiner (One Plus One). In *Lawrence Weiner: As Far As The Eye Can See*, Ann Goldstein and Donna De Salvo, ed. pp. 320–355. The Museum of Contemporary Art, Los Angeles, and the Whitney Museum of American Art, New York.

Collier, Mike. 2013. On Ways of Walking and Making Art: A Personal Reflection. In *Walk On: From Richard Long to Janet Cardiff: 40 Years of Art Walking*, Mike Collier and Cynthia Morrison-Bell, ed. (with assistance from Janet Ross), 72–81. Sunderland, UK: Art Editions North/University of Sunderland.

Collier, Mike, and Cynthia Morrison-Bell, ed. 2013. *Walk On: From Richard Long to Janet Cardiff: 40 Years of Art Walking*. Sunderland, UK: Art Editions North/University of Sunderland.

Cottee, Simon, and Keith Hayward. 2011. Terrorist (E)motives: The Existential Attractions of Terrorism. *Studies in Conflict and Terrorism*, 34(12): 963–986. https://doi.org/10.1080/1057610X.2011.621116.

Cotter, Holland. 2009. The Collected Ingredients of a Beijing Life. *The New York Times*, July 15: C1, C5. https://www.nytimes.com/2009/07/15/arts/design/15song.html.

De Salvo, Donna. 2007. As Far As the Eye Can See. In *Lawrence Weiner: As Far As The Eye Can See*, Ann Goldstein and Donna De Salvo, ed. pp. 58–99. The Museum of Contemporary Art, Los Angeles, and the Whitney Museum of American Art, New York.

Dutton, Denis. 2009. Has Conceptual Art Jumped the Shark Tank. *The New York Times* (New York edition), October 16: A27. Published online as "Has Conceptual Art Jumped the Shark Tank" on October 15, 2009, at https://www.nytimes.com/2009/10/16/opinion/16dutton.html.

Eastwood, Lauren E. 2006. Contesting the Economic Order and Media Construction of Reality. In *Igniting a Revolution: Voices in Defense of the Earth*, Steven Best and Anthony J. Nocella, II, ed. pp. 114–126. Oakland, CA: AK Press.

Farrer, Ben, and Graig R. Klein. 2022. How Radical Environmental Sabotage Impacts US Elections. *Terrorism and Political Violence*, 34(2): 218–239. https://doi.org/10.1080/09546553.2019.1678468.

Fineberg, Jonathan. 1995. *Art Since 1940: Strategies of Being*. Englewood Cliffs, N.J.: Prentice Hall.

Fox, Margalit. 2015. Chris Burden, 69, a Conceptualist With Scars, Dies. *The New York Times* (New York edition), May 12: B14. Published online as "Chris Burden, a Conceptualist With Scars, Dies at 69," on May 11, 2015, at https://www.nytimes.com/2015/05/12/arts/chris-burden-a-conceptualist-with-scars-dies-at-69.html.

Fuchs, R.H. 1986. *Richard Long*. London and New York: Thames and Hudson Ltd./The Solomon R. Guggenheim Foundation.

Fulton, Hamish. 1991. Old Muddy. In *Richard Long: Walking in Circles*, 241–247. New York: George Braziller.

Ghorashi, Hannah. 2015. The New York Times Releases a Short Documentary On The Man Who Shot Chris Burden. *ARTNews*, May 20. https://www.artnews.com/art-news/news/the-new-york-times-releases-a-short-documentary-on-the-man-who-shot-chris-burden-4206/.

Gillen, Eckhart. 2004. The Artist as Perpetrator. There can be no Creation without Destruction. In *legal/illegal: Art beyond Law*, Corinna Weidner, ed. pp. 131–144. Berlin: NGBK.

Gillick, Liam. 2007. The Objectification of Desire in the Work of Lawrence Weiner (Finding a Place in the Sun). In *Lawrence Weiner: As Far As The Eye Can See*, Ann Goldstein and Donna De Salvo, ed. pp. 242–275. The Museum of Contemporary Art, Los Angeles, and the Whitney Museum of American Art, New York.

Godfrey, Tony. 1998. *Conceptual Art*. London: Phaidon.

Goldstein, Ann. 2007. If It Looks Like a Duck and It Walks Like a Duck, It Is Probably a Duck. In *Lawrence Weiner: As Far As The Eye Can See*, Ann Goldstein and Donna De Salvo, ed. pp. 100–159. The Museum of Contemporary Art, Los Angeles, and the Whitney Museum of American Art, New York. New Haven, CT: Yale University Press.

Hayward, Keith J. 2004. *City Limits: Crime, Consumer Culture and the Urban Experience*. London: Glasshouse Press.

Hayward, Keith, and Marc Schuilenberg. 2014. To resist = to create? Some thoughts on the concept of resistance in cultural criminology. *Tijdschrift over Cultuur & Criminaliteit*, 4(1): 22–36. https://doi.org/10.5553/TCC/221195072013003003003.

Herbert, Martin. 2021. How Lawrence Weiner (1942–2021) Reimagined What Art Could Be. *ArtReview*, December 10. https://artreview.com/how-lawrence-weiner-1942-2021-reimagined-what-art-could-be/.

Higgins, Charlotte. 2012. Richard Long: 'It was the swinging 60s. To be walking lines in the fields was a bit different'. *The Guardian*, June 15. https://www.theguardian.com/artanddesign/2012/jun/15/richard-long-swinging-60s-interview.

Hoberman, J. 2004. Pop and Circumstance. *The Nation*, November 24. https://archive.ph/20050301174151/http://www.thenation.com/doc.mhtml?i=20041213&s=hoberman&c=1.

Hopkins, David. 2000. *After Modern Art: 1945–2000*. Oxford: Oxford University Press.

Illing, Sean. 2021. The Case for a More Radical Climate Movement: Author Andreas Malm on the Failures of Climate Activism and the Need for Escalation. *Vox*, October 1. https://www.vox.com/vox-conversations-podcast/22691428/vox-conversations-climate-change-andreas-malm.

Jaworowski, Ken. 2017. Burden. *The New York Times* (New York edition), May 5: C8. Published online as "Review: In 'Burden,' an Artist Who Is Tortured. Literally." on May 4, 2017, at https://www.nytimes.com/2017/05/04/movies/burden-review-chris-burden.html.

Jones, Jonathan. 2009a. Richard Long takes art for a walk at Tate Britain. *The Guardian*, May 27. https://www.theguardian.com/artanddesign/jonathanjonesblog/2009/may/27/richard-long-tate-britain-exhibition.

Jones, Jonathan. 2009b. A hymn of love to the earth. *The Guardian*, June 2. https://www.theguardian.com/artanddesign/2009/jun/03/richard-long-exhibition-tate.

Jordan, John. 1998. The Art of Necessity: The Subversive Imagination of Anti-road Protest and Reclaim the Streets. In *DiY Culture: Party & Protest in Nineties Britain*, George McKay, ed. pp. 129–151. London and New York: Verso.

Kaye, Nick. 2018. Liveness and the entanglement with things. In *Artists in the Archive: Creative and Curatorial Engagements with Documents of Art and Performance*, Paul Clarke, Simon Jones, Nick Kaye, and Johanna Linsley, ed. pp. 25–51. Abingdon, Oxon, UK, and New York: Routledge.

Kennedy, Randy. 2007. Language as Sculpture, Words as Clay. *The New York Times*, October 21: AR31. https://www.nytimes.com/2007/10/21/arts/design/21kenn.html.

Kennedy, Randy. 2013. The Balance of a Career. *The New York Times* (New York edition), September 8: AR75. Published online as "The Balance of a Career" on September 6, 2013, at https://www.nytimes.com/2013/09/08/arts/design/chris-burdens-feats-of-art-are-to-fill-the-new-museum.html.

Kirby, Michael. 1994. Foreword. In *Conceptual Art: An American Perspective*, Robert C. Morgan, pp. ix–xi. Jefferson, NC, and London: McFarland & Company, Inc., Publishers.

Kutner, Eric. 2015. 'Shot in the Name of Art'. *The New York Times*, May 20. https://www.nytimes.com/2015/05/20/opinion/shot-in-the-name-of-art.html.

Lambert, Jonathan. 2019. Does Empathy Have a Dark Side? *National Public Radio* (*N.P.R.*), April 12. https://www.npr.org/sections/health-shots/2019/04/12/712682406/does-empathy-have-a-dark-side.

Leffingwell, Edward. 2007. Ships at Sea. In *Lawrence Weiner: As Far As The Eye Can See*, Ann Goldstein and Donna De Salvo, ed. pp. 276–319. The Museum of Contemporary Art, Los Angeles, and the Whitney Museum of American Art, New York.

Lennard, Natasha. 2019. How the Prosecution of Animal Rights Activists as Terrorists Foretold Today's Criminalization of Dissent. *The Intercept*, December 12. https://theintercept.com/2019/12/12/animal-people-documentary-shac-protest-terrorism/.

Lentricchia, Frank, and Jody McAuliffe. 2003. *Crimes of Art and Terror*. Chicago, IL: University of Chicago Press.

LeWitt, Sol. 1967. Paragraphs on Conceptual Art. *Artforum*, 5(10) [June/Summer]: 79–83. https://www.artforum.com/features/paragraphs-on-conceptual-art-211354/ (last visited October 26, 2024; on file with author).

LeWitt, Sol. [1969] 1972. Sentences on Conceptual Art, 1968. In *Conceptual Art*, Ursula Meyer, pp. 174–179. New York: E.P. Dutton. Reprinted from *Art-Language*, Vol. 1, No. 1 (1969).

Lippard, Lucy R., ed. [1973] 1997. *Six Years: The Dematerialization of the Art Object*. Berkley and Los Angeles, CA: University of California Press.

Loadenthal, Michael. 2017. "Eco-Terrorism": An Incident-Driven History of Attack (1973–2010). *Journal for the Study of Radicalism*, 11(2) [Fall]: 1–34. https://www.jstor.org/stable/26297171.

Long, Richard. 1991. Fragments of a Conversation V [with Anne Seymour]. In *Richard Long: Walking in Circles*, pp. 92–93. New York: George Braziller.

Long, Richard. [1988] 1991. An Interview with Richard Long by Richard Clark. In *Richard Long: Walking in Circles*, pp. 248–252. New York: George Braziller. [Broadcast in 'Third Ear', by BBC Radio 3, Friday 28 October 1988. Produced by Judith Bumpus.]

Long, Richard. [1997] 2002. A Transcription of Richard Long's Talk During A Slide Show Given At The Naoshima Contemporary Art Museum, Japan, On May 25 1997. In *Richard Long: Walking the Line*, pp. 146–147. London and New York: Thames & Hudson Ltd/Thames & Hudson Inc.

Long, Richard. [2008] 2009. And So Here We Are: A Conversation with Michael Craig-Martin, November 2008. In *Richard Long: Heaven and Earth*, Clarrie Wallis, ed. pp. 172–177. London: Tate Publishing.

Lorimer, Hayden. 2016. Walking: New Forms and Spaces for Studies of Pedestrianism. In *Geographies of Mobilities: Practices, Spaces, Subjects*, Tim Cresswell and Peter Merriman, ed. pp. 19–33. Abingdon, Oxon, UK, and New York: Routledge.

Macfarlane, Robert. 2009. Walk the line. *The Guardian*, May 22. https://www.theguardian.com/artanddesign/2009/may/23/richard-long-photography-tate-britain.

Malm, Andreas. 2021. *How to Blow Up a Pipeline: Learning to Fight in a World on Fire*. London and New York: Verso.

Marchese, David. 2024. How This Climate Activist Justifies Political Violence. *The New York Times*, January 16. https://www.nytimes.com/interactive/2024/01/14/magazine/andreas-malm-interview.html.

Meikle, Graham. 2002. *Future Active: Media Activism and the Internet*. New York and London: Routledge.

Meyer, Ursula. 1972. Introduction. In *Conceptual Art*, Ursula Meyer, pp. VII–XX. New York: E.P. Dutton.

Moorhouse, Paul. 2002. The Intricacy of the Skein, The Complexity of the Web: Richard Long's Art. In *Richard Long: Walking the Line*, pp. 29–43. London and New York: Thames & Hudson Ltd/Thames & Hudson Inc.

Morgan, Robert C. 1994. *Conceptual Art: An American Perspective*. Jefferson, NC, and London: McFarland & Company, Inc., Publishers.

Morgan, Robert C. 1996. *Art Into Ideas: Essays on Conceptual Art*. Cambridge: Cambridge University Press.

Morris, Blake. 2020. *Walking Networks: The Development of an Artistic Medium*. London and New York: Rowman & Littlefield.

Morrison-Bell, Cynthia. 2013. Foreword. In *Walk On: From Richard Long to Janet Cardiff: 40 Years of Art Walking*, Mike Collier and Cynthia Morrison-Bell, ed. (with assistance from Janet Ross), pp. 1–3. Sunderland, UK: Art Editions North/University of Sunderland.

O'Dell, Kathy. 1998. *Contract with the Skin: Masochism Performance Art and the 1970s*. Minneapolis, MN: University of Minnesota Press.

O'Hagan, Sean. 2009. One step beyond. *The Guardian*, May 9. https://www.theguardian.com/artanddesign/2009/may/10/art-richard-long.

O'Rourke, Karen. 2013. *Walking and Mapping: Artists as Cartographers*. Cambridge, MA, and London: The MIT Press.

Osborne, Peter. 1999. Conceptual Art and/as Philosophy. In *Rewriting Conceptual Art*, Michael Newman and Jon Bird, ed. pp. 47–65. London: Reaktion Books.

Phillpot, Clive. 1998. Books by Artists and Books As Art. In *Artist/Author: Contemporary Artists' Books*, Cornelia Lauf and Clive Phillpot, ed. pp. 30–55. New York: Distributed Art Publishers Inc., and The American Federation of Arts.

Pickering, Leslie James. 2007. *The Earth Liberation Front, 1997–2002*, 2nd edn. Portland, OR: Arissa Media Group.

Plagens, Peter. 1973. He Got Shot—For His Art. *The New York Times*. September 2: D1, D3 (on file with author).

Resistance Magazine. 2010. Pot Calling the Kettle Black: 60 Years of USA brand terrorism vs. 14 years of earth liberation direct action. *Resistance, Journal of the Earth Liberation Movement*, 12–14 (on file with author).

Robinson, Alistair. 2013. On Walking. In *Walk On: From Richard Long to Janet Cardiff: 40 Years of Art Walking*, Mike Collier and Cynthia Morrison-Bell, ed. (with assistance from Janet Ross), pp. 14–21. Sunderland, UK: Art Editions North/University of Sunderland.

Rorimer, Anne. 1999. Siting the Page: Exhibiting Works in Publications—Some Examples of Conceptual Art in the USA. In *Rewriting Conceptual Art*, Michael Newman and Jon Bird, ed. pp. 11–26. London: Reaktion Books.

Rosenberg, Karen. 2009. Amsterdam as Hub for Globetrotting Conceptualists. *The New York Times*, July 21: C5. https://www.nytimes.com/2009/07/21/arts/design/21modern.html.

Rosin, Hanna. 2019. The End of Empathy. *National Public Radio* (*N.P.R.*), April 15. https://www.npr.org/2019/04/15/712249664/the-end-of-empathy.

Salter, Colin. 2011. Activism as Terrorism: The Green Scare, Radical Environmentalism and Governmentality. *Anarchist Developments in Cultural Studies*, No. 1(2011): Ten Years After 9/11: An Anarchist Evaluation: 211–238. https://journals.uvic.ca/index.php/adcs/article/view/17126/7340.

Schwarz, Dieter. 1989. *Lawrence Weiner, Books 1968–1989: Catalogue Raisonné*. Köln: Verlag der Buchhandlung Walther König/Villeurbanne: Le Nouveau Musée.

Schwarz, Dieter. 2007. The Metaphor Problem, Again and Again: Books and Other Things by Lawrence Weiner. In *Lawrence Weiner: As Far As The Eye Can See*, Ann Goldstein and Donna De Salvo, ed. pp. 160–207. The Museum of Contemporary Art, Los Angeles, and the Whitney Museum of American Art, New York. New Haven, CT: Yale University Press.

Serota, Nicholas. 2009. Walking Abroad. In *Richard Long: Heaven and Earth*, Clarrie Wallis, ed. pp. 28–29. London: Tate Publishing.

Seymour, Anne. 1991. Walking in Circles. In *Richard Long: Walking in Circles*, pp. 7–39. New York: George Braziller.

Seymour, Anne. 2002. Foreword. In *Richard Long: Walking the Line*, pp. 7–10. London and New York: Thames & Hudson Ltd/Thames & Hudson Inc.

Smith, Rebecca K. 2008. "Ecoterrorism"?: A Critical Analysis of the Vilification of Radical Environmental Activists as Terrorists. *Environmental Law*, 38(2) [Spring]: 537–576. https://law.lclark.edu/live/files/17299-38-2smith.

Smith, Phil. 2010. The Contemporary Derive: A Partial Review of Issues Concerning the Contemporary Practice of Psychogeography. *Cultural Geographies*, 17(1): 103–122. https://doi.org/10.1177/1474474009350002.

Smith, William. 2018. Disruptive Democracy: The Ethics of Direct Action. *Raisons Politiques*, 2018/1 (69): 13–27. https://shs.cairn.info/journal-raisons-politiques-2018-1-page-13?lang=en.

Stephens-Griffin, Nathan. 2023. Biting Back: A Green-Cultural Criminology of Animal Liberation Struggle as Constructed Through Online Communiques. *Crime Media Culture*, 19(2): 252–273. https://doi.org/10.1177/17416590221110118.

Stockhausen, Karlheinz. 2001. Message from Professor Karlheinz Stockhausen. September 19. http://www.stockhausen.org/message_from_karlheinz.html (last visited September 9, 2009).

Strick, Jeremy, and Adam D. Weinberg. 2007. Director's Foreword. In *Lawrence Weiner: As Far As The Eye Can See*, Ann Goldstein and Donna De Salvo, ed. pp. 38–41. The Museum of Contemporary Art, Los Angeles, and the Whitney Museum of American Art, New York. New Haven, CT: Yale University Press.

Taylor, Brandon. 2009. Absurd lines, protest walks: notes on Richard Long. *Sculpture Journal*, 18(2): 176–188. https://doi.org/10.3828/sj.2009.5;

https://authoringobjects.wordpress.com/wp-content/uploads/2016/01/absurd_lines__protest_walks__n.pdf.

Tommasini, Anthony. 2001. Music; The Devil Made Him Do It. *The New York Times* (National edition), September 30: §2: 28. Published online as "Music; The Devil Made Him Do It" on September 30, 2001, at https://www.nytimes.com/2001/09/30/arts/music-the-devil-made-him-do-it.html.

Tufnell, Ben, ed. 2007. *Richard Long: Selected Statements & Interviews*. London: Haunch of Venison.

"Interview with William Furlong (1984)" (pp. 59–63). *From an interview given in London in February 1984. Originally published as an audio tape by* Audio Arts, *London, 1985.*

"Interview with Martina Giezen (1985–1986)" (pp. 65–91). *From interviews conducted in Bristol on 19 November 1985, London on 7 April 1986, and Amsterdam on 13 April 1986. Originally published in two parts as* Richard Long In Conversation, *MW Press, Noordwijk, Holland, 1986.*

"Interview with Colin Kirkpatrick (1994)" (pp. 97–103). *Interview given on 8 July 1994 on the occasion of a solo exhibition at the Pier Arts Centre, Stromness, Orkney.*

"Interview with Yuko Hasegawa (1995)" (pp. 105–107). *Interview given 12 October 1995. Originally published as 'Questions for Richard Long' in* Sango Suigyo, *Setagaya Art Museum, Tokyo, 1996.*

"Interview with Michael Auping (2000)" (pp. 111–119). *Interview given in the Quinton House pub, on 3 June, 2000, on the occasion of a solo exhibition at the Royal West of England Academy, Bristol. To be published in Michael Auping,* 30 Years: Interviews and Outtakes, *Modern Art Museum of Fort Worth in association with Prestel, 2007.*

Vanderheiden, Steve. 2005. Eco-terrorism or Justified Resistance? Radical Environmentalism and the "War on Terror." *Politics & Society*, 33(3): 425–447. https://doi.org/10.1177/0032329205278462.

Vetrocq, Marcia E. 1999. Conceptualism: An Expanded View. *Art in America*, 87(7) [July]: 72–77.

Vogel, Carol. 2009. Subway Riders Are Greeted by a Blast of Sol LeWitt Color. *The New York Times*, September 14: C1, C7. Published online as "Subway Riders Are Greeted by a Blast of Sol LeWitt Color" on September 13, 2009, at https://www.nytimes.com/2009/09/14/arts/design/14lewitt.html.

Wallis, Clarrie. 2009. Making Tracks. In *Richard Long: Heaven and Earth*, Clarrie Wallis, ed. pp. 33–61. London: Tate Publishing.

Weiner, Lawrence. [1969] 1972. LAWRENCE WEINER, October 12, 1969 [In conversation with the editor.] In *Conceptual Art*, Ursula Meyer, pp. 217–218. New York: E.P. Dutton.

Weiner, Lawrence. 1998. *Lawrence Weiner*. London and New York: Phaidon.

Wilson, Andrew. 2009. From Page to Page: An introduction to Richard Long's books. In *Richard Long: Heaven and Earth*, Clarrie Wallis, ed. pp. 172–177. London: Tate Publishing.

Wood, William. 1999. Still You Ask for More: Demand, Display and 'The New Art.' In *Rewriting Conceptual Art*, Michael Newman and Jon Bird, ed. pp. 66–87. London: Reaktion Books.

CHAPTER 4

Conclusion

As ecodefense became more appealing as a tactic in the years following the publication of *The Monkey Wrench Gang*, first-hand accounts of participation in and reasons for direct action to slow or halt development projects, safeguard natural resources, and protect the environment (including animal species) began to emerge (e.g., Foreman 1991; Foreman and Haywood 1993; Hill 2000; Hunter 1979; Manes 1990; Mann 2009; Watson 1981, 1993, 1996, 2003), as did in-depth examinations of the people, actions, histories, and philosophies behind "radical environmentalism" (e.g., Hochschartner 2017; Morris 1995; Newkirk 2000; Rosebraugh 2004; Scarce 1990, 2006; Zakin 1993). Around the same time—and certainly since then—scholarly literature (has) analyzed and assessed, ***inter alia***, the asserted justifications for ecodefense (see, e.g., Cianchi 2013, 2015; Hays et al. 1996; Manson 2024; Martin 1990: 298–309; Shevory 1996; Smith 2018; Vanderheiden 2005), the legislative responses to E.L.F. and E.L.F.-like tactics (see, e.g., Amster 2006; Hummell 2004; Karasick 2009; Moore 2005; see also Brisman 2008: 758–759; Ellefsen 2016), and the ethical dilemmas faced by researchers studying "radical environmentalists" (see, e.g., Scarce 1990, 1994, 1995, 2005, 2006).[1] Scholars have also considered the appropriate moniker for illegal activities undertaken in defense of the Earth—whether to refer to the phenomena of "ecodefense," "ecotage," "monkeywrenching," etc.,

A. Brisman, *Direct Action as Conceptual Art?*, Palgrave Studies in Crime, Media and Culture,
https://doi.org/10.1007/978-3-031-76634-3_4

as "ecoterrorism" (or "eco-terrorism")[2]—or whether "states and corporations [are] 'the real eco-terrorists'" (Pellow 2020: 566)[3]—as well as the morality of breaking the law for the purpose of ecological protection and whether economic sabotage, property destruction, and related tactics even constitute "violence."[4]

Although radical criminologies and critical criminologies, as a whole, have often been more concerned with investigating the meaning of criminal behavior than in trying to isolate specific causes (see, e.g., Brisman 2014; Muncie 2013)—and cultural criminology, in particular, has been devoted to disentangling the elaborate web through which the meaning of crime and crime control is created, constructed, enforced, reinforced, and resisted[5]—and although green cultural criminology has focused attention on representations of and resistance to environmental crimes and harms (see discussion in Chapter 1)—there has been little criminological consideration of the *meaning* of direct action to protect the environment (but see Cianchi 2013, 2015; Joosse 2007, 2012, 2021; and Stephens-Griffin 2023 for rare exceptions). While a handful of studies have analyzed *media coverage* of the A.L.F., the E.L.F., and "radical environmental resistance" (see, e.g., Dawn 2004; Eastwood 2006; Joosse 2012; Laurendeau and Van Brunschot 2007)[6]—and notwithstanding excellent qualitative content analysis by Johnston and Johnston (2017, 2020) and Stephens-Griffin (2023)—there has been relatively little criminological research and scholarship devoted to the significance of the communiqué.[7]

Direct Action as Conceptual Art?: An Examination of the Role of the Communiqué for Ecodefense has attempted to understand some of the meaning of ecodefense (conceived of here to include direct action in defense of or on behalf of the environment and nonhuman animals (e.g., economic sabotage, property destruction, lab raids, "live liberations") by examining the communiqué—the written statement or press release often issued after the action. By juxtaposing the communiqué with written statements and documents by Conceptual artists, this book has tried to demonstrate that ecodefense is not a singular endeavor or set of activities reducible to criminological considerations of causation, prevention, and control. In addition, because "[f]ew... concepts are as loaded with ideology or coloured by a particular moment" than "terrorism" (Malm 2021: 107), this book has endeavored to extract ecodefense from the muddled mass of "terrorist" jargon (see Allhoff and Buciak 2013; Finlay 2009; Joosse 2012; see also Berry 2012: 237; Ellefsen 2012: 201; see generally Pickering 2007: 54–56; Yates 2007: 140–141, 145, 147, 148–51)—a task consistent with cultural criminology's goal of understanding the meaning of crime, integral to critical criminology's commitment to critiquing and opposing (systems of) domination, and vital to green cultural criminology's interest in the representation of environmental harm and resistance thereto—not to mention one (still) *necessary* in the post-9/11 "eco-paranoia" atmosphere (Allhoff and Buciak 2013: 225).

According to Ellefsen (2016: 443, 441), political repression—"a specific type of social control whose purpose is to prevent or diminish direct and non-institutional challenges to social, cultural and/or political power, i.e. protest, activism and social movements (Earl 2011, p. 262)"—is an important factor in "determining the opportunities and constraints existing during any [social] movement's life span." A little more than ten years ago, some suggested that because of political repression—because many of the more active A.L.F. and E.L.F. cells had been quashed in the United States (in what has been referred to as the "Green Scare"[8])—and because European "eco-radicals" had begun operating with a lower profile—ecodefense (as I have broadly construed it) had become a less relevant form of resistance (see, e.g., Allhoff and Buciak 2013: 236; Federman 2013; Holden 2011: C5; see generally Peachey 2014; cf. Cianchi 2013, 2015; Taylor 2013). Writing at the beginning of the second decade of the twenty-first century, Allhoff and Buciak (2013: 238) concluded that "[t]oday, radical environmentalism seems to be obsolescent due to state repression, lack of charismatic characters and ineffective public relations. That Osama Bin Laden warned of climate change as an outcome of Western imperialism does not help; nor does the fact that Ted Kaczynski is sometimes considered an eco-terrorist for his anti-technology beliefs."[9] But environmental degradation, loss of biodiversity, harm to nonhuman animals, and despoliation of the Earth, as a whole, are not diminishing; we *are* facing a crisis and the situation is getting much worse. With growing awareness of climate change and increased manifestations of its negative effects—leading some to refer to it as a "climate emergency" (see, e.g., Lee 2021: 157, 159, 160, 166; Todd 2019; but see Hulme 2019), "radical environmentalists" may well (re)turn to ecodefense (see Brown 2021: 371; Malm 2021; see also Illing 2021; Schlossberg 2021; Wickenden 2021; Wolfe 2022/2023; see generally Curra 2014: 77).

More than a decade ago, Allhoff and Buciak (2013: 238) suggested that climate change is "a multi-causal, relatively diffuse phenomenon... [that] does not offer clear adversaries to act against." Climate change remains "multi-causal" insofar as different industries or sectors continue to be responsible for greenhouse gas emissions (e.g., agriculture, manufacturing, power (fossil fuel), transportation), but it cannot—and *should not*—be considered a "diffuse phenomenon": the increased frequency, extent, and severity of wildfires from hotter temperatures and extended drought have blistered, cracked, and scarred the Earth; more severe storms and warmer, rising oceans have flooded and inundated, eroded and stained low-lying islands and coastal countries. The evidence is everywhere. Climate change is also not dispersed or geographically isolated, nor is it achromatopsic: it is color blind, in that it affects us all regardless of skin color, but those impacts have been and will continue to be distributed unevenly and unequally, and various groups have been and will continue to be in different positions to adapt to climate change (Brisman 2015). Moreover, as many have suggested—with Andreas Malm perhaps the most recent and most prominent—climate change no longer lacks "clear adversaries to act against"—assuming it ever did.

Malm (2021: 155) claims that "the climate movement took off *because* it had no connections to the ecosystem of [Earth First!], ALF and ELF" (emphasis in original). Had the climate movement "started with ecotage," he insists, "it would have gone nowhere" (2021: 155). "All those thousands of monkeywrenching actions," he maintains, "achieved little if anything and had no lasting gains to show for them" (2021: 155). "They were not performed in a dynamic relation to a mass movement," Malm (2021: 155) asserts, "but largely in a void." Whether Malm's analysis here is correct or unnecessarily captious is largely irrelevant. Should we encounter acts of ecodefense (or ones that appear as such)—whether in response to Malm's call for escalation or independent of the climate movement—the communiqué may well take on heightened significance.

Notes

1. Scarce, who was conducting research on environmental protest, was jailed for contempt of court because he declined to tell a federal grand jury what, if anything, he knew about various acts of ecodefense. For a discussion of the ethical dilemma Scarce faced regarding

whether to speak to law enforcement authorities, see Monaghan (1993); Witt (2009: 37, 2011: 41).

2. See, e.g., Allhoff and Buciak (2013: 224–26); Amster (2006: 288, 289–294, 297, 299–300); Brisman (2008: 756–757); Brown (2020: 240 (citing Asal and Rethemeyer 2008; Taylor 1998)); Chapter 1 n.9 and 11 and accompanying text; Gleick (2006: 483–84); Hall and Farrall (2013: 131n.9 and accompanying text); Johnston and Johnston (2020: 570–571, 575–576 (citing Cooke 2013; Hirschmann 2000; Humphrey and Stears 2006; Liddick 2006; Monaghan 1999, 2013; Sorenson 2009; Vanderheiden 2005)); Joosse (2014: 35 (quoting Vanderheiden 2008)); Loadenthal (2017: 2, 4, 7, 15, 25); Mancuso-Smith (2006); Martin (1990); Smith (2008); see also Di Ronco (2023: 16, 17); Di Ronco and Chiaramonte (2022: 422); Hasler et al. (2020: 520, 526, 528); Manson (2024: 3); McIntyre (2021: 9–12); Pellow (2020: 566); Salter (2011: 212, 218, 219, 226, 227, 228, 229, 230, 232); Selmini and Di Ronco (2023: 205, 207, 219, 220, 221); Vanderheiden (2005: 426, 447, 441, 445); see generally Stephens-Griffin (2023: 260–263, 269); Woodhouse (2018: 223).

   In her chapter, "Contesting the Economic Order and Media Construction of Reality," Eastwood (2006: 114, 121) puts forth the argument that "[n]ot only have the connections between the environment, politics, and consumption been minimized, glossed over, or simply made invisible, but [that] activities and ideologies which have the potential to contest the current economic order are likewise significantly diffused by the media"—a dynamic she finds unsurprising because of "the media's collusion with capital accumulation in this current phase of globalization." "Rather than making the connections between hyper-consumption, environmental degradation, and human and animal rights violations clear," she continues, "the popular media engages in several tactics which serve to maintain the invisibility of the connections between the way our current economic order is structured and the environmental and human casualties of this order" (2006: 121–122). One of these tactics, Eastwood (2006: 122) explains, involves "represent[ing] those individuals who contest the economic order as 'terrorist'…."

   Similarly, Selmini and Di Ronco (2023: 205) conclude that "the criminalization of political dissent is done not only by enacting and enforcing criminal (or subcriminal) law but also through use of public and media discourses that successfully construct social activists as a problem." In addition, Selmini and Di Ronco (2023: 205) remind us, "[t]his framing strategy also pursues symbolic violence: it not only depoliticizes and delegitimizes social movements but also legitimizes punitive responses against them and their mobilization."

Whereas Eastwood's chapter (2006) centers on the relationship between corporations and "the popular media" (Selmini and Di Ronco's (2023) article spotlights the criminalization of dissent and protest in Western liberal democracies)—and it is worth bearing in mind that the concentration of media ownership is even more pronounced in the mid-2020s (twenty years after Eastwood's writing)—Watts (2016) focuses on the role of the state in the construction of "terrorism" and "terrorist." Watts (2016: 117) suggests that states (get to) define what is—and what is *not*—"terrorism," reminding us of "the power of the state to represent a problem in a particular way (Bacchi 1999, 2009)." As Watts (2016: 115) explains, "[p]re-eminent among the powers possessed by states is the capacity to shape the symbolic world and where the meaning of power means the power to mean. Political and legal discourses are understood not as merely cognitive or contemplative entity but as an articulatory and performative practice which constitutes and organizes normal and legitimate social relations and practices through the use of constitutive categories." "At the same time," Watts (2016: 115) continues, "the state possesses and uses the same symbolic power to declare exceptions to the norm enabling it to step beyond what is normal and lawful"—an insight that foreshadows the author's later book, *Criminalizing Dissent: The Liberal State and the Problem of Legitimacy* (2020).

3. Amster (2006) would certainly concur with this perspective. As he argues, "[i]n this historical moment... [of] surging environmental assault... the true ecoterrorism is precisely what it sounds like: terrorizing the environment. Taking this view, the targets confronted by ALF and ELF would actually be the terrorists, since they generally involve animal research, genetic modification of plants, and the promotion of ecologically unsound practices such as driving SUVs" (2006: 287, 296).

Ruggiero (2017: 313) makes a similar point, albeit in the broader context of institutional violence ("violence from above") and anti-institutional violence ("violence from below"): "there exists a monopoly of sorts in the definition of behaviour, and before defining a behaviour as 'terrorist', definitions must be filtered through the discourses of official politics and the media." As Ruggiero (2017: 313) explains, "[m]anifestations of political dissent, extreme though they may be, should be seen against social and institutional processes, tentatively exemplified by Ferracuti (1982: 130) in the following words: 'Terrorism could be defined as what the other person does. What we do, or the state, do is anti-terrorism. But obviously the positions can be reversed by shifting side[s], or simply the flow of history.'" For Ruggiero (2017: 313–314), "[a] definition incorporating such social and institutional processes might be: a terrorist act is any act carried out during the course of political struggle, aimed at influencing, conquering or *defending* state power, implying the use of extreme violence against innocent, non-combatant persons. This definition includes both institutional and anti-institutional violence, namely acts against a state and acts by states against their internal or external enemies" (emphasis in original). Therefore, Ruggiero (2017: 314) continues, "terrorism" constitutes "a form of 'pure' violence exercised by organized forces inflicting mass violence on civilians. It is random: it does not target specific actors whose conduct is regarded as wrongful, but general populations which are hit because of their nationality, ethnicity, religious or political creed. Terrorism adopts a concept of collective liability against its victims.... [C]ontemporary wars waged by Western countries contain all of these definitional elements."

Finlay (2009) considers this issue in depth, although, like Ruggiero (2017), his focus is on "terrorism," more broadly, rather than "ecoterrorism," in particular. Finlay (2009: 755, 752) acknowledges that "terrorism" and "terrorist" are "commonly used with reference to non-state entities rather than states" (footnote omitted), and he quotes Scheffler (2006: 1) for the proposition that "[t]he term 'terrorism' may by now have become too ideologically freighted to have any analytic value. If the term is to be an aid to understanding, two opposed but complementary ways of employing it will have to be resisted. On the one hand, there is the tendency, among the representatives and defenders of governments facing violent threats from non-state groups and organizations, to use the term to refer to all forms of political violence perpetrated by non-state actors. On the other hand, there is the tendency, among the representatives and defenders of non-state actors engaged in political violence, to insist that 'the real terrorists' are the officials or the military forces of those states with which they are locked in conflict."

In their section on "Animal innocents and 'the real criminals'," Johnston and Johnston (2020: 569) note that "[a]lthough all RALM activists strive for justice through the liberation of non-human animals, not all feel that liberation equals justice for all *human* animals." Indeed, as Johnston and Johnston (2020: 569–570) illuminate, "[t]he most significant example of dissonance within RALM framings of justice was the question of whether prison is the rightful place for 'real criminals' or whether all forms of imprisonment should be abolished." Later, in the same section, Johnston and Johnston (2020: 570) illustrate how the documents they examined exposed "a clear underlying message about the injustice of a legal system that would sanction the torture and murder of animals while prosecuting those who damage inanimate objects as an act of political resistance." "These sentiments," they continue (2020: 570), "reflect a paradox in the leftist position that, on the one hand, is critical of tough-on-crime politics that result in greater reliance on incarceration, but simultaneously demands punitive carceral responses towards actors within the animal industry."

4. See, e.g., Best and Nocella (2004a); Brown (2021); Martin (1990); Regan (2004); Young (2001); Yourofsky (2010); see also Aaltola (2012: 157, 165); Allhoff and Buciak (2013: 222–224); Amster (2006: 289); Foreman (1991: 113–15); Johnston and Johnston (2017: 737); Loadenthal (2017: 3–7, 9, 20–21, 23–25); Manson (2024: 2–6, 10–13, 15); Rosebraugh (2004: 91); Shevory (1996: 190); Smith (2018: 16, 18); see generally Finlay (2009: 773 (quoting Scheffler 2006: 3)); Leech (2012: 155); McIntyre (2021: 14); Vanderheiden (2005: 427).

   Malm (2021: 102) chooses to concede that "property destruction *is* violence, insofar as it intentionally exerts physical force to inflict injury on a thing owned by someone who does not want it to happen" (emphasis added). But by the same token, Malm (2021: 102) clarifies, "we must insist on it being *different in kind* from the violence that hits a human (or an animal) in the face, for the reasons just specified: one cannot treat a car cruelly or make it cry. . . . [O]ne cannot equate the treatment of people with the treatment of the things they own" (emphasis in original). Malm (2021: 103) then identifies one exception—one instance—in which "property destruction . . . approaches killing and maiming, namely that which hits material conditions for subsistence: poisoning someone's groundwater, burning down a family's last remaining grove of olive trees or, for that matter, firebombing a paddy field in an Indian peasant village because it emits methane." This, Malm (2021: 103) contends, "would come close to a stab in the heart." But "[a]t the other end of the spectrum," Malm (2021: 103) asserts without equivocation, "is the blasting of a superyacht into smithereens." Essentially, Malm (quoted in Illling 2021) draws a distinction "between slashing a tire and slashing the lungs of the owner of the car. . . . I accept th[e] philosophical definition and commonsense use of the term . . that property destruction is a form of violence. But it is a lesser form of violence, qualitatively different from harming human beings."

For broader discussions on the meaning and representation of violence, see, e.g., Stanko (2003); (Žižek 2008); for a consideration of whether a revolution (always) entails violence, see Hayward and Schuilenberg (2014: 30n.3).

5. Cultural criminology emerged and began to take shape in the mid-1990s, and literature on, about, and exemplifying (or "doing") cultural criminology since then has been substantial. For formative statements and orienting perspectives, see, e.g., Barak (1994); Brisman and South (2013, 2014); Ferrell (1999, 2013); Ferrell et al. (2004); Ferrell, Hayward and Young (2015); Ferrell and Sanders (1995); Hayward (2011, 2016); Hayward and Ilan (2017); Hayward and Young (2004); Ilan (2019); Manning (1998); Young (2008).
6. For an analysis of "terrorists'" use of the media, see Ross (2007: 216–17); for the marginalization of counter-hegemonic groups by the mainstream media, see Joosse (2012: 76, 90); for reflections on how the E.L.F. engaged with the mainstream media (primarily between the years of 1997 and 2002), see Pickering (2007: 6–8).
7. Stephens-Griffin's (2023: 253) employed qualitative content analysis of animal liberation communiqués published on the *Bite Back* website between January 1, 2020, and December 31, 2020, "to examine the way animal liberation activists represent and construct their actions in defence of non-human animals." Johnston and

Johnston (2017: 738 (endnote omitted), 2020: 567) conducted qualitative content analysis of "North American RALM activist websites and public Facebook pages dating from 2005-2015"—documents comprised of communiqués, essays, Facebook posts, published interviews, mission statements, and pamphlets (2017: 738, 2020: 567)—to study how activism within the RALM intersects with other social movements (in the case of their 2017 article) and to explore how RALM activists perceive justice (their 2020 article). In an endnote, Johnston and Johnston (2017: 748n.3) clarify that "[c]ommuniqués made up the bulk of the ALF activist data," which they describe as "they anonymous statements issued by activists to 'press officers' who publish them on the activist website." Consistent with my statements about the functions and purposes of communiqués in Chapter 2, Johnston and Johnston (2017: 748n.3) proffer that "[c]ommuniqués inform readers of the direct actions claimed by autonomous cells of the ALF, offer a way of seeing how activists conceptualize their activism as it occurs in the present moment rather than in retrospect (Liddick 2006), and provide the most instructive method of determining paradigmatic impacts in clandestine movements where activists cannot be identified and studied directly because of the potential illegality or covert nature of their actions (Joosse 2014)."

8. See, e.g., Federman (2013); Potter (2010, 2011); Rosenfeld (2006); Salter (2011); see also Brown (2019); Hasler et al. (2020); Pellow (2020); Wolfe (2022/2023); cf. Resistance Magazine (2009). Brown (2020: 249, 248) contends that the E.L.F.'s period of "most intense activity" occurred between "the late 1990s through the mid-2000s," but that "[a]s government forces cracked down on the ELF's aboveground press office [the NAELFPO]" and as "the ELF's underground cells were targeted by a massive FBI investigation ('Operation Backfire') which led to a series of arrests in 2005 and 2006," the number of actions/attacks declined.

   In outlining the historical precedents for the "Green Scare," Salter (2011: 221) asserts that "[t]here is a direct parallel between the use of the term *domestic terrorism* today with the Red Scare of the 1940s and 1950s. The term *terrorist* has today replaced *communist* as the foremost rhetorical wedge for contemporary Western states. The selective and asymmetric targeting of individuals, groups and groupings indicate that 'green' has effectively become the new 'red' (Potter 2011)" (emphasis in original; internal footnote omitted).

9. At the time of Allhoff and Buciak's writing (2013), Theodore John "Ted" Kaczynski, the so-called "Unabomber," was serving consecutive life sentences in the United States Penitentiary, Administrative Maximum Facility (USP Florence ADMAX) (also known as "ADX Florence" and the "Florence Supermax") in Florence, Colorado. In late 2021, Kaczynski was transferred to Federal Medical Center, Butner (FMC Butner), an administrative security medical center in Butner, North Carolina—a facility that treats incarcerated individuals suffering from significant health problems (Sisak et al. 2023; Zapotsky 2021; see also Thrush 2023; Traub 2023). Kaczynski was found unresponsive in his cell on June 10, 2023, and his death was later ruled a suicide (Sisak et al. 2023; Thrush 2023; see also Loskutoff 2024).

   In contrast to Allhoff and Buciak (2013), Semley and Millar (2021), in an article in *The Baffler* (an American magazine of art, cultural analysis, and political criticism) written prior to Kaczynski's transfer to FMC Butner and subsequent death, reflected on how "the Unabomber's pop culture stock ha[d] been rising." Citing Semley and Millar (2021), Traub (2023), in his obituary of Kaczynski in *The New York Times*, noted that Kaczynski's ideas had generated online support and had inspired "eco-terrorists," although he (Traub), admitted that the "online support did not indicate how many eco-terrorists had been newly minted . . . ." For a note about Kaczynski with respect to "green politics," see Manson (2024: 4).

References

Aaltola, Elisa. 2012. Differing Philosophies: Criminalisation and the Stop Huntington Animal Cruelty Debate. In *Eco-global Crimes: Contemporary Problems and Future Challenges*, Rune Ellefsen, Ragnhild Sollund, and Guri Larsen, ed. pp. 157–180. Surrey, UK: Ashgate.

Allhoff, Steffen W., and Sebastian K. Buciak. 2013. Action, Reaction, Escalation: Radical Environmental Organisations in the Context of Eco-terrorism and Eco-radicalism. In *Global Environmental Change: New Drivers for Resistance, Crime and Terrorism?*, Achim Maas, Balázs Bodó, Clementine Burnley, Irina Comardicea, and Roger Roffey, ed. pp. 221–242. Baden-Baden, DE: Nomos Verlagsgesellschaft.

Amster, Randall. 2006. Perspectives on Ecoterrorism: Catalysts, Conflations, and Casualties. *Contemporary Justice Review*, 9(3): 287–301. https://doi.org/10.1080/10282580600827991.

Asal, Victor H., and R. Karl Rethemeyer. 2008. Dilettantes, Ideologues, and the Weak: Terrorists Who Don't Kill. *Conflict Management and Peace Science*, 25(3): 244–263. https://doi.org/10.1080/07388940802219000.

Bacchi, Carol Lee. 1999. *Women, Policy and Politics: The Construction of Policy Problems*. London and Thousand Oaks, CA: Sage.

Bacchi, Carol Lee. 2009. *Analysing Policy: What's the Problem Represented to Be?* Frenchs Forest, NSW: Pearson Australia.

Barak, Gregg. 1994. *Media, Process, and the Social Construction of Crime: Studies in Newsmaking Criminology*. New York: Garland.

Berry, Bonnie. 2012. A Critical Interpretation of Animal Exploitation. In *Routledge Handbook of Critical Criminology*, Walter S. DeKeseredy and Molly Dragiewicz, ed. pp. 230–240. London and New York: Routledge.

Best, Steven, and Anthony J. Nocella, II. 2004a. Introduction: Behind the Mask: Uncovering the Animal Liberation Front. In *Terrorists or Freedom Fighters?: Reflections on the Liberation of Animals*, Steven Best and Anthony J. Nocella, II, ed. pp. 9–63. New York: Lantern Books.

Brisman, Avi. 2008. Crime-Environment Relationships and Environmental Justice. *Seattle Journal for Social Justice*, 6(2): 727–817. https://digitalcommons.law.seattleu.edu/cgi/viewcontent.cgi?article=1138&context=sjsj.

Brisman, Avi. 2014. Of Theory and Meaning in Green Criminology. *International Journal for Crime, Justice and Social Democracy*, 3(2): 22–35. https://doi.org/10.5204/ijcjsd.v3i2.173; https://www.crimejusticejournal.com/article/view/723.

Brisman, Avi. 2015. 'Multicolored' Green Criminology and Climate Change's Achromatopsia. *Contemporary Justice Review*, 18(2): 178–196. https://doi.org/10.1080/10282580.2015.1025629.

Brisman, Avi, and Nigel South. 2013. A Green-Cultural Criminology: An Exploratory Outline. *Crime Media Culture*, 9(2): 115–135. https://doi.org/10.1177/1741659012467026.

Brisman, Avi, and Nigel South. 2014. *Green Cultural Criminology: Constructions of Environmental Harm, Consumerism, and Resistance to Ecocide*. Abingdon, Oxon, UK, and New York: Routledge.

Brown, Alleen. 2019. The Green Scare: How a Movement That Never Killed Anyone Became the FBI's No. 1 Domestic Terrorism Threat. *The Intercept*, March 23. https://theintercept.com/2019/03/23/ecoterrorism-fbi-animal-rights/.

Brown, Joseph M. 2020. Notes to the Underground: Credit Claiming and Organizing in the Earth Liberation Front. *Terrorism and Political Violence*, 32(2): 237–256. https://doi.org/10.1080/09546553.2017.1364637. Published online on September 7, 2017.

Brown, Joseph M. 2021. Civil Disobedience, Sabotage, and Violence in US Environmental Activism. In *The Oxford Handbook of Comparative Environmental Politics*, Jeannie Sowers, Stacy D. VanDeveer, and Erika Weinthal, ed. pp. 356–374. Online edition, Oxford Academic, at https://doi.org/10.1093/oxfordhb/9780197515037.013.34.

Cianchi, John. 2013. *I talked to my tree and my tree talked back: radical environmental activists and their relationships with nature*. Ph.D. Thesis, School of Social Sciences, University of Tasmania (Australia).

Cianchi, John. 2015. *Radical Environmentalism: Nature, Identity and More-than-human Agency*. Basingstoke, Hampshire, UK: Palgrave Macmillan.

Cooke, Steve. 2013. Animal rights and environmental terrorism. *Journal of Terrorism Research*, 4(2): 26–36. https://doi.org/10.15664/jtr.532; https://research-repository.st-andrews.ac.uk/bitstream/handle/10023/4898/532-1525-1-PB.pdf?sequence=1&isAllowed=y.

Curra, John. 2014. *The Relativity of Deviance*, 3rd edn. Thousand Oaks, CA, and London: Sage.

Dawn, Karen. 2004. From the Front Line to the Front Page—An Analysis of ALF Media Coverage. In *Terrorists or Freedom Fighters?: Reflections on the Liberation of Animals*, Steven Best and Anthony J. Nocella II, ed. pp. 213–228. New York: Lantern Books.

Di Ronco, Anna. 2023. *Policing Environmental Protest: Power and Resistance in Pandemic Times*. Bristol, UK: Bristol University Press.

Di Ronco, Anna, and Xenia Chiaramonte. 2022. Harm to Knowledge: Criminalising Environmental Movements Speaking Up Against Megaprojects. In *The Palgrave Handbook of Restorative Justice*, Brunilda Pali, Miranda Fosyth, and Felicity Tepper, ed. pp. 421–447. Cham, CH: Palgrave Macmillan.

Earl, Jennifer. 2011. Political Repression: Iron Fists, Velvet Gloves, and Diffuse Control. *Annual Review of Sociology*, 37: 261–284. https://doi.org/10.1146/annurev.soc.012809.102609.

Eastwood, Lauren E. 2006. Contesting the Economic Order and Media Construction of Reality. In *Igniting a Revolution: Voices in Defense of the Earth*, Steven Best and Anthony J. Nocella, II. ed. pp. 114–126. Oakland, CA: AK Press.

Ellefsen, Rune. 2012. Green Movements as Threats to Order and Economy: Animal Activists Repressed in Austria and Beyond. In *Eco-global Crimes:*

Contemporary Problems and Future Challenges, Rune Ellefsen, Ragnhild Sollund, and Guri Larsen, ed. pp. 181–205. Surrey, UK: Ashgate.

Ellefsen, Rune. 2016. Judicial Opportunities and the Death of SHAC: Legal Repression Along a Cycle of Contention. *Social Movement Studies*, 15(5): 441–456. https://doi.org/10.1080/14742837.2016.1185360.

Federman, Adam. 2013. Climate Activists, Eco-Terrorism, and the Green Scare. *Utne*, September/October. http://www.utne.com/politics/climate-activists-zm0z13sozlin.aspx#axzz3N9uska3M (last visited August 15, 2024; on file with author).

Ferracuti, Franco. 1982. A Sociopsychiatric Interpretation of Terrorism. *The ANNALS of the American Academy of Political and Social Science*, 463(1): 129–140. https://doi.org/10.1177/000271628246300 1011; https://www-jstor-org.libproxy.eku.edu/stable/1043617.

Ferrell, Jeff. 1999. Cultural Criminology. *Annual Review of Sociology*, 25: 395–418.

Ferrell, Jeff. 2013. Cultural Criminology. In *The Sage Dictionary of Criminology*, 3rd edn. Eugene McLaughlin and John Muncie, ed. pp. 110–112. London and Thousand Oaks, CA: Sage.

Ferrell, Jeff, Keith Hayward, Wayne Morrison, and Mike Presdee, ed. 2004. *Cultural Criminology Unleashed*. London: Cavendish.

Ferrell, Jeff, Keith Hayward, and Jock Young. 2015. *Cultural Criminology: An Invitation*, 2nd edn. London: Sage.

Ferrell, Jeff, and Clinton R. Sanders, ed. 1995. *Cultural Criminology*. Boston: Northeastern University Press.

Finlay, Christopher J. 2009. How to do Things with the word 'Terrorist.' *Review of International Studies*, 35(4) [October]: 751–774. http://www.jstor.org/stable/40588072.

Foreman, Dave. 1991. *Confessions of an Eco-Warrior*. New York: Crown.

Foreman, Dave, and Bill Haywood, ed. 1993. *Ecodefense: A Field Guide to Monkeywrenching*, 3rd ed. Chico, CA: Abbzug Press.

Gleick, Peter H. 2006. Water and Terrorism. *Water Policy*, 8(6): 481–503. https://doi.org/10.2166/wp.2006.035.

Hall, Matthew, and Stephen Farrall. 2013. The Criminogenic Consequences of Climate Change: Blurring the Boundaries Between Offenders and Victims. In *Routledge International Handbook of Green Criminology*, Nigel South and Avi Brisman, ed. pp. 120–133. Abingdon, Oxon, UK, and New York: Routledge.

Hasler, Olivia, Reece Walters, and Rob White. 2020. In and Against the State: The Dynamics of Environmental Activism. *Critical Criminology: An International Journal*, 28(3): 517–531. https://doi.org/10.1007/s10612-019-094 32-0.

Hays, Scott, Michael Esler, and Carol Hays. 1996. Radical Environmentalism and Crime. In *Environmental Crime and Criminality: Theoretical and Practical Issues*, Sally M. Edwards, Terry D. Edwards, and Charles B. Fields, ed. pp. 163–181. New York and London: Garland Publishing.

Hayward, Keith. 2011. The Critical Terrorism Studies-Cultural Criminology Nexus: Some Thoughts on how to 'Toughen Up' the Critical Studies Approach. *Critical Studies on Terrorism*, 4(1): 57–73. https://doi.org/10.1080/17539153.2011.553387.

Hayward, Keith J. 2016. Cultural Criminology: Script Rewrites. *Theoretical Criminology*, 20(3): 297–321. https://doi.org/10.1177/1362480615619668.

Hayward, Keith, and Jonathan Ilan. 2017. In *The Routledge Companion to Criminological Theory and Concepts*, Avi Brisman, Eamonn Carrabine and Nigel South, ed. pp. 262–266. Abingdon, Oxon, UK, and New York: Routledge.

Hayward, Keith, and Marc Schuilenberg. 2014. To Resist = To Create? Some Thoughts on the Concept of Resistance in Cultural Criminology. *Tijdschrift over Cultuur & Criminaliteit*, 4(1): 22–36. https://doi.org/10.5553/TCC/221195072013003003003.

Hayward, Keith J., and Jock Young. 2004. Cultural Criminology: Some Notes on the Script. *Theoretical Criminology*, 8(3) [August]: 259–273. https://doi.org/10.1177/1362480604044608.

Hill, Julia Butterfly. 2000. *The Legacy of Luna: The Story of a Tree, a Woman and the Struggle to Save the Redwoods*. San Francisco, CA: Harper San Francisco.

Hirschmann, Kai. 2000. The changing face of terrorism. *Internationale Politik und Gesellschaft* [*International Politics and Society*], 3/2000: 299–310. https://www.fes.de/ipg/ipg3_2000/arthirschmannneu.htm.

Hochschartner, Jon. 2017. *The Animals' Freedom Fighter: A Biography of Ronnie Lee, Founder of the Animal Liberation Front*. Jefferson, NC: McFarland & Company Inc.

Holden, Stephen. 2011. Crimes Against Property, as Protests. *The New York Times* (New York edition), June 22: C5. Published online as "Crimes Against Property, as Protests" on June 21, 2011, at https://www.nytimes.com/2011/06/22/movies/if-a-tree-falls-documentary-on-earth-liberation-front.html?searchResultPosition=1.

Hulme, Mike. 2019. Climate Emergency Politics Is Dangerous. *Issues in Science and Technology*, 36(1) [Fall]: 23–25. https://www.jstor.org/stable/26949073.

Hummell, Katrina Albright. 2004. The American Legislative Exchange Council's Attack on Civil Liberties: A Critical Analysis of the "Animal and Ecological Terrorism Act and "Environmental Corrupt Organizations-Preventative Legislation and Neutralization Act." http://noaeta.org/LegalResources/AETA.LegalAnalysisDOW.pdf (last visited December 12, 2014; on file with author).

Humphrey, Matthew, and Marc Stears. 2006. Animal Rights Protest and the Challenge to Deliberative Democracy. *Economy and Society*, 35(3): 400–422. https://doi.org/10.1080/03085140600844969.

Hunter, Robert. 1979. *Warriors of the Rainbow: A Chronicle of the Greenpeace Movement*. New York: Holt, Rinehart and Winston.

Ilan, Jonathan. 2019. Cultural Criminology: The Time is Now. *Critical Criminology: An International Journal*, 27(1): 5–20. https://doi.org/10.1007/s10612-019-09430-2.

Illing, Sean. 2021. The Case for a More Radical Climate Movement: Author Andreas Malm on the Failures of Climate Activism and the Need for Escalation. *Vox*, October 1. https://www.vox.com/vox-conversations-podcast/22691428/vox-conversations-climate-change-andreas-malm.

Johnston, Genevieve, and Matthew S. Johnston. 2017. 'We Fight for all Living Things': Countering Misconceptions About the Radical Animal Liberation Movement. *Social Movement Studies*, 16(6): 735–751. https://doi.org/10.1080/14742837.2017.1319268.

Johnston, Genevieve, and Matthew S. Johnston. 2020. 'Until Every Cage is Empty': Frames of Justice in the Radical Animal Liberation Movement. *Contemporary Justice Review*, 23(4): 563–580. https://doi.org/10.1080/10282580.2020.1719361.

Joosse, Paul. 2007. Leaderless Resistance and Ideological Inclusion: The Case of the Earth Liberation Front. *Terrorism and Political Violence*, 19(3): 351–368. https://doi.org/10.1080/09546550701424042.

Joosse, Paul. 2012. Elves, Environmentalism, and "Eco-terror": Leaderless Resistance and Media Coverage of the Earth Liberation Front. *Crime Media Culture*, 8(1): 75–93. https://doi.org/10.1177/1741659011433366.

Joosse, Paul. 2014. Antiglobalization and Radical Environmentalism: An Exchange on Ethical Grounds. *Ethics in Progress*, 5(1): 33–51. https://doi.org/10.14746/eip.2014.1.2; https://pressto.amu.edu.pl/index.php/eip/article/view/9382/9084.

Joosse, Paul. 2021. Narratives of Rebellion. *European Journal of Criminology*, 18(5): 735–754. https://doi.org/10.1177/147737081987442.

Karasick, Paul J. 2009. Curb Your Ecoterrorism: Identifying the Nexus Between State Criminalization of Ecoterror and Environmental Protection Policy. *William and Mary Environmental Law and Policy Review*, 33(2): 581–603. https://scholarship.law.wm.edu/cgi/viewcontent.cgi?referer=&httpsredir=1&article=1037&context=wmelpr.

Laurendeau, Jason, and Erin Gibbs Van. Brunschot. 2007. Home-Grown Brand of Al Qaeda': Moral Panics and the Construction of 'Eco-terrorism. *International Journal of Crime, Criminal Justice and Law*, 2(1): 105–123.

Lee, Murray. 2021. Policing the Pedal Rebels: A Case Study of Environmental Activism Under COVID-19. *International Journal for Crime, Justice and*

Social Democracy, 10(2): 156–168. https://doi.org/10.5204/ijcjsd.1887; https://www.crimejusticejournal.com/article/view/1887/1069.

Leech, Garry. 2012. *Capitalism: A Structural Genocide*. London and New York: Zed Books.

Liddick, Donald R. 2006. *Eco-Terrorism: Radical Environmental and Animal Liberation*. Westport, CT: Praeger.

Loadenthal, Michael. 2017. "Eco-Terrorism": An Incident-Driven History of Attack (1973–2010). *Journal for the Study of Radicalism*, 11(2) [Fall]: 1–34. https://www.jstor.org/stable/26297171.

Loskutoff, Maxim. 2024. The Poisoned Dream of the American West. *The New York Times* (New York edition), June 16: SR10. Published online as "The Unabomber, Me and the Poisoned Myth of the American West" on June 9, 2024, at https://www.nytimes.com/2024/06/09/opinion/unabomber-novel-american-west.html?searchResultPosition=2.

Malm, Andreas. 2021. *How to Blow Up a Pipeline: Learning to Fight in a World on Fire*. London and New York: Verso.

Mancuso-Smith, Chrystal. 2006. From Monkeywrenching to Mass Destruction: Eco-Sabotage and the American West. *Journal of Land, Resources, and Environmental Law*, 26: 319–338.

Manes, Christopher. 1990. *Green Rage: Radical Environmentalism and the Unmaking of Civilization*. Boston: Little, Brown and Co.

Mann, Keith. 2009. *From Dusk 'til Dawn: An insider's view of the growth of the Animal Liberation Movement*. London: Warcry Communications.

Manning, Peter K. 1998. Media Loops. In *Popular Culture, Crime, and Justice*, Frankie Y. Bailey and Donna C. Hale, ed. pp. 25–39. Belmont, CA: West/Wadsworth.

Manson, Dylan. 2024. Eco-sabotage as Defensive Activism. *Ethical Theory and Moral Practice*. 27(4) [September]: 505–522. https://doi.org/10.1007/s10677-024-10449-w.

Martin, Michael. 1990. Ecosabotage and Civil Disobedience. *Environmental Ethics*, 12(4): 291–310.

McIntyre, Iain. 2021. *Environmental Blockades: Obstructive Direct Action and the History of the Environmental Movement*. Abingdon, Oxon, UK, and New York: Routledge.

Monaghan, Peter. 1993. Sociologist Jailed Because He 'Wouldn't Snitch' Ponders the Way Research Ought to Be Done. *The Chronicle of Higher Education*, 40 (September 1): A8, A9.

Monaghan, Rachel. 1999. Terrorism in the Name of Animal Rights. *Terrorism and Political Violence*, 11(4): 159–169. https://doi.org/10.1080/09546559908427538.

Monaghan, Rachel. 2013. Not Quite Terrorism: Animal Rights Extremism in the United Kingdom. *Studies in Conflict & Terrorism*, 36(11): 933–951. https://doi.org/10.1080/1057610X.2013.832117.

Moore, Andrew N. Ireland. 2005. Caging Animal Advocates' Political Freedoms: The Unconstitutionality of the Animal and Ecological Terrorism Act. *Animal Law*, 11(1): 255–282. https://lawcommons.lclark.edu/alr/vol11/iss1/10.

Morris, David B. 1995. *Earth Warrior: Overboard with Paul Watson and the Sea Shepherd Conservation Society*. Golden, CO: Fulcrum Publishing.

Muncie, John. 2013. Radical Criminologies. In *The Sage Dictionary of Criminology*, 3rd edn. Eugene McLaughlin and John Muncie, ed. pp. 361. London and Thousand Oaks, CA: Sage.

Newkirk, Ingrid. 2000. *Free the Animals: The Amazing Trust Story of the Animal Liberation Front*. New York: Lantern Books.

Peachey, Paul. 2014. Animal rights group ends 15-year campaign against experiments at Huntingdon. *The Independent*, August 24. https://www.independent.co.uk/news/uk/crime/animal-rights-group-ends-15year-campaign-against-experiments-at-huntingdon-9687843.html.

Pellow, David N. 2020. Environmental justice, animal rights and total liberation From conflict and distance to points of common focus In *Routledge International Handbook of Green Criminology*, 2nd edn. Avi Brisman and Nigel South, ed. pp. 555–570. Abingdon, Oxon, UK, and New York: Routledge.

Pickering, Leslie James. 2007. *The Earth Liberation Front, 1997–2002*, 2nd edn. Portland, OR: Arissa Media Group.

Potter, Will. 2010. While Everyone Goes Green Uncle Sam Goes for the Throat. *Resistance, Journal of the Earth Liberation Movement*, Spring: 24–26 (on file with author).

Potter, Will. 2011. *Green is the New Red: An Insider's Account of a Social Movement Under Siege*. San Francisco, CA: City Lights Books.

Regan, Tom. 2004. *How to Justify Violence*. In *Terrorists or Freedom Fighters?: Reflections on the Liberation of Animals*, Steven Best and Anthony J. Nocella, II, ed. pp. 231–236. New York: Lantern Books.

Resistance Magazine. 2009. On the Frontlines: With Captain Paul Watson and the Sea Shepherd Conservation Society. *Resistance, Journal of the Earth Liberation Movement*, Fall: 14–19, 31, 35 (on file with author).

Rosebraugh, Craig. 2004. *Burning Rage of a Dying Planet: Speaking for the Earth Liberation Front*. New York: Lantern Books.

Rosenfeld, Ben. 2006. The Great Green Scare and the Fed's "Case" Against Rod Coronado. *CounterPunch*, March 10. https://www.counterpunch.org/2006/03/10/the-great-green-scare-and-the-fed-s-quot-case-quot-against-rod-coronado/.

Ross, Jeffrey Ian. 2007. Deconstructing the Terrorism-News Media Relationship. *Crime Media Culture*, 3(2): 215–225. https://doi.org/10.1177/1741659007078555.

Ruggiero, Vincenzo. 2017. Institutional and Anti-institutional violence. In *The Routledge Companion to Criminological Theory and Concepts*, Avi Brisman, Eamonn Carrabine and Nigel South, ed. pp. 311–315. Abingdon, Oxon, UK, and New York: Routledge.

Salter, Colin. 2011. Activism as Terrorism: The Green Scare, Radical Environmentalism and Governmentality. *Anarchist Developments in Cultural Studies*, No. 1(2011): Ten Years After 9/11: An Anarchist Evaluation: 211–238. https://journals.uvic.ca/index.php/adcs/article/view/17126/7340.

Scarce, Rik. 1990. *Eco-Warriors: Understanding the Radical Environmental Movement*, Chicago, IL: Noble Press.

Scarce, Rik. 2006. *Eco-Warriors: Understanding the Radical Environmental Movement*, Updated Edition. Walnut Creek, CA: Left Coast Press.

Scarce, Rik. 1994. (No) Trial (But) Tribulations: When Courts and Ethnography Conflict. *Journal of Contemporary Ethnography*, 23(July): 123–149.

Scarce, Rik. 1995. Scholarly Ethics and Courtroom Antics: Where Researchers Stand in the Eyes of the Law. *American Sociologist*, 26(Spring): 87–112.

Scarce, Rik. 2005. A Law to Protect Scholars. *The Chronicle of Higher Education*, (August 12): 324. https://www.chronicle.com/article/a-law-to-protect-scholars/.

Scheffler, Samuel. 2006. Is Terrorism Morally Defensible? *Journal of Political Philosophy*, 14(1) [March]: 1–17. https://doi.org/10.1111/j.1467-9760.2006.00242.x; https://as.nyu.edu/content/dam/nyu-as/philosophy/documents/faculty-documents/scheffler/Scheffler-Samuel-Is-Terrorism-Morally-Distinctive.pdf.

Schlossberg, Tatiana. 2021. Climate Disaster. *The New York Times Book Review*, January 24: 22. Published online as "Three Books Offer New Ways to Think About Environmental Disaster" on January 22, 2021, at https://www.nytimes.com/2021/01/22/books/review/scorched-earth-emmanuel-kreike-how-to-prepare-for-climate-change-david-pogue-how-to-blow-up-a-pipeline-andreas-malm.html.

Selmini, Rossella, and Anna Di Ronco. 2023. The Criminalization of Dissent and Protest. *Crime and Justice*, 52(1): 197–231. https://doi.org/10.1086/727553.

Semley, John, and Edward Millar. 2021. Influencer Society and Its Future: Swallow the Ted pill on Unabomber stan TikTok. *The Baffler*, September 27. https://thebaffler.com/latest/influencer-society-and-its-future-semley-millar.

Shevory, Thomas C. 1996. Monkeywrenching: Practice in Search of a Theory. In *Environmental Crime and Criminality: Theoretical and Practical Issues*, Sally

M. Edwards, Terry D. Edwards, and Charles B. Fields, ed. pp. 183–204. New York and London: Garland Publishing.

Sisak, Michael R., Mike Balsamo, and Jake Offenhartz. 2023. 'Unabomber' Ted Kaczynski died by suicide in prison medical center, AP sources say. *Associated Press*, June 11. https://apnews.com/article/ted-kaczynski-unabomber-1197f597364b36e56bdbcaca9837bdc4.

Smith, Rebecca K. 2008. "Ecoterrorism"?: A Critical Analysis of the Vilification of Radical Environmental Activists as Terrorists. *Environmental Law*, 38(2) [Spring]: 537–576. https://law.lclark.edu/live/files/17299-38-2smith.

Smith, William. 2018. Disruptive Democracy: The Ethics of Direct Action. *Raisons Politiques*, 2018/1 (69): 13–27. https://shs.cairn.info/journal-raisons-politiques-2018-1-page-13?lang=en.

Sorenson, John. 2009. Constructing terrorists: propaganda about animal rights. *Critical Studies on Terrorism*, 2(2): 237–256. https://doi.org/10.1080/17539150903010715.

Stanko, Elizabeth A., ed. 2003. *The Meanings of Violence*. New York and London: Routledge.

Stephens-Griffin, Nathan. 2023. Biting Back: A Green-Cultural Criminology of Animal Liberation Struggle as Constructed Through Online Communiques. *Crime Media Culture*, 19(2): 252–273. https://doi.org/10.1177/17416590221110118.

Taylor, Bron. 1998. Religion, Violence, and Radical Environmentalism: From Earth First! to the Unabomber to the Earth Liberation Front. *Terrorism and Political Violence*, 10(4): 1–42. https://doi.org/10.1080/09546559808427480.

Taylor, Bron. 2013. Resistance: Do the Ends Justify the Means? *Earth Island Journal*, April 26. https://www.earthisland.org/journal/index.php/articles/entry/resistance_do_the_ends_justify_the_means/.

Thrush, Glenn. 2023. Kaczynski Died by Suicide, Prompting Questions of Prison Security. *The New York Times*, June 11: A20. Published online as "Kaczynski Is Said to Have Died by Suicide in Prison" on June 10, 2023, at https://www.nytimes.com/2023/06/10/us/politics/kaczynski-unabomber-suicide-prison.html.

Todd, Matthew. 2019. The Climate Emergency and the End of Diversity. In *This Is Not a Drill: An Extinction Rebellion Handbook*, Clare Farrell, Alison Green, Sam Knights, and William Skeaping, ed. pp. 69–72. Milton Keynes, UK: Penguin/Random House.

Traub, Alex. 2023. Theodore J. Kaczynski, Boy Genius Turned 'Unabomber,' Dies at 81. *The New York Times* (New York edition), June 12: B5. Published online as "Ted Kaczynski, 'Unabomber' Who Attacked Modern Life, Dies at 81," on June 10, 2023 [updated June 11, 2023], at https://www.nytimes.com/2023/06/10/us/ted-kaczynski-dead.html.

Vanderheiden, Steve. 2005. Eco-terrorism or Justified Resistance? Radical Environmentalism and the "War on Terror." *Politics & Society*, 33(3): 425–447. https://doi.org/10.1177/0032329205278462.

Vanderheiden, Steve. 2008. Radical environmentalism in an age of antiterrorism. *Environmental Politics*, 17(2) [April]: 299–318. https://doi.org/10.1080/09644010801936248; https://sciencepolicy.colorado.edu/admin/publication_files/2008.47.pdf.

Watson, Paul. 1981. *Sea Shepherd: My Fight for Whales and Seals*. New York: W.W. Norton.

Watson, Paul. 1993. *Earthforce: An Earth Warrior's Guide to Strategy*. Los Angeles, CA: Chaco Press.

Watson, Paul. 1996. *Ocean Warrior: My Battle to End the Illegal Slaughter on the High Seas*. Toronto, ON: Key Porter Books.

Watson, Paul. 2003. *Seal Wars: Twenty-Five Years on the Front Lines with the Harp Seals*. Buffalo, NY: Firefly Books.

Watts, Rob. 2016. *States of Violence and the Civilising Process: On Criminology and State Crime*. London: Palgrave Macmillan/Springer Nature.

Watts, Rob. 2020. *Criminalizing Dissent: The Liberal State and the Problem of Legitimacy*. Abingdon, Oxon, UK, and New York: Routledge.

Wickenden, Dorothy. 2021. Andreas Malm on the Environmental Movement and "Intelligent Sabotage." *The New Yorker*, September 27. https://www.newyorker.com/podcast/political-scene/andreas-malm-on-the-environmental-movement-and-intelligent-sabotage.

Witt, Jon. 2009. *SOC*. New York: McGraw-Hill.

Witt, Jon. 2011. *SOC*. New York: McGraw-Hill.

Wolfe, Matthew. 2022/2023. Backfire. *The New York Times Magazine*, May 29: 38. Published online as "The Rise and Fall of America's Environmentalist Underground" on May 26, 2022 [updated June 15, 2023], at https://www.nytimes.com/2022/05/26/magazine/earth-liberation-front-joseph-mahmoud-dibee.html.

Woodhouse, Keith Makato. 2018. *The Ecocentrists: A History of Radical Environmentalism*. New York: Columbia University Press.

Yates, Roger. 2007. Debating 'Animal Rights' Online: The Movement-Countermovement Dialectic Revisited. In *Issues in Green Criminology: Confronting harms against environments, humanity and other animals*, Piers Beirne and Nigel South, ed. pp. 140–157. Cullompton, Devon, UK: Willan.

Young, Alison. 2008. Culture, Critical Criminology and the Imagination of Crime. In *The Critical Criminology Companion*, Thalia Anthony and Chris Cunneen, ed. pp. 18–29. Leichhardt, NSW: Hawkins Press/The Federation Press.

Young, Thomas. 2001. The Morality of Ecosabotage. *Environmental Values*, 10(3) [August]: 385–393. https://www.jstor.org/stable/30301826.

Yourofsky, Gary. 2010. Box 2.1 Abolition, Liberation, Freedom. In *Essential Criminology*, 3rd edn. Mark M. Lanier and Stuart Henry, ed. pp. 27–31. Boulder, CO: Westview Press.

Zakin, Susan. 1993. *Coyotes and Town Dogs: Earth First! and the Environmental Movement*. New York: Viking.

Zapotosky, Matt. 2021. Ted Kaczynski, the 79-year-old Unabomber, Transferred to Prison Medical Facility. *The Washington Post*, December 22. https://www.washingtonpost.com/national-security/unabomber-kaczynski-hospital-ill/2021/12/22/4f773f82-6367-11ec-8ce3-9454d0b46d42_story.html.

Žižek, Slavoj. 2008. *Violence: Six Sideways Reflections*. New York: Picador.

Index

A. Brisman, *Direct Action as Conceptual Art?*, Palgrave Studies in Crime, Media and Culture,
https://doi.org/10.1007/978-3-031-76634-3

H

I

T

V

W

The manufacturer's authorised representative in the EU is Springer Nature Customer Service Centre GmbH, Europaplatz 3, 69115 Heidelberg, Germany. If you have any concerns regarding our products, please contact ProductSafety@springernature.com

Printed and bound by CPI Group (UK) Ltd, Croydon, CR0 4YY
01/10/2026
02225903-0001